Chicken Coops for the SOUL

With love to my daughters

Chicken Coops for the Soul

A henkeeper's story

JULIA HOLLANDER

guardianbooks

Published by Guardian Books 2011

2 4 6 8 10 9 7 5 3 1

First published in Great Britain in 2010 by
Guardian Books
Kings Place, 90 York Way
London N1 9GU

www.guardianbookshop.co.uk

A CIP catalogue record for this book is available from the British Library

ISBN 9780852652350

Typeset by seagulls.net

Printed and bound in Great Britain by CPI Bookmarque, Croydon

'You must have heard of chickens; they're all the rage!'

Spike Milligan

Contents

Funky Chicken

Those who spend a night with a chicken
Will cackle in the morning.
Tunisian proverb

I confess that I, like too many others before me, became a chicken-owner on a whim. It was five years ago and those few weeks before Christmas when presents urgently needed finalising. My eldest daughter Ellie was insisting that her one and only wish was for a bunny rabbit; she had been saying so since the previous Christmas. My objections revolved around the usual – you are not old enough; I know it's going to be me that has to look after it; I'm not sure keeping a pet in a cage is kind, especially without company … each time I came up with a fresh and valid argument, it crashed against a wall of five-year-old willpower.

A major weakness was that I couldn't remember how old I had been when I myself got a rabbit. It was possible I too had only just started school, was only just able to take responsibility for cleaning my teeth, let alone keeping another little creature alive. What

I did remember was the huge satisfaction of caring for Ravioli the rabbit: building him a cosy nest; the feeding and the watering; the delights of both cuddling and taunting him; the drama of protecting him from the ravenous family dog. Later Ravioli turned out to be a mere warm-up for my even greater dedication to ponies.

How was it possible that since starting a family, I had survived this long without any animals at all?

On a drizzly Saturday morning Ellie and I made our way through chaotic traffic, out into open countryside. I was happy that this shopping trip was taking us away from the dazzle and bluster of the crowds, happy not to be jostling over plastic toys and computer games.

We were heading towards the animal sanctuary where as a toddler Ellie had first clasped a real live bunny. I remembered the occasion well – the two of them perfectly matched in warmth and softness.

As we entered we spotted them, there in the big cage – a pair of bunnies snuggled against the wire. White with black points, they were; Himalayans; small enough for even a five year old to handle. And behind them, skittering around the logs and stones, was a matching pair of chickens. The combination was irresistible.

'You don't want them useless bantams,' said the old lady who ran the sanctuary, leaving Ellie to bond with the bunnies and taking me round the back of the building. 'What you need is a couple of Warrens.' Warren sounded like a nice, traditional breed. Though my first thought had been to provide the rabbit with a feathered companion or two, already I was moving into a more utilitarian mode – here was a pet that in return for its care might provide us with breakfast. I was mighty impressed when, from the flock of identical red hens in her shed, she selected the two that would lay the best; through long experience, I thought, she must have developed some profound intuition about their abilities.

By the time we returned, Ellie had chosen her Himalayan and christened him – Snowy. As the sanctuary lady lowered each of the Warrens into an empty sack, stapling shut the opening in case they tried to escape, Ellie begged for the opportunity to name them too. I wasn't sure. One of the few things I knew about chickens was that they were extremely attractive to foxes, and round our terraced town house we have foxes galore. Surely their snitching a hen with a name would feel much worse than one without. Quite rightly, Ellie argued that I had let her name her bunny even though the fox might try to steal him. So I gave in and let her call them 'Roxy' and 'Loxy'. She had no explanation for her choices, except that they rhymed with 'Foxy'.

Then slowly home we drove, a two-storey hutch strapped to the roof; the sacks of chicken wedged behind my seat and a couple more of feed on the passenger seat. In the back, Ellie sat bolt upright with Snowy in a box on her lap, a frown of solicitude on her face. This was the beginning of a whole new era.

'How long do you think the hens might live?' I had enquired as we departed.

'Ooh, ages,' the lady at the sanctuary assured me. '11,12 years, if you look after them.'

Great, I thought – that's like a dog or a cat; Roxy and Loxy could just about be around until Ellie leaves home. She will be off, forging a life for herself, and meanwhile maybe I shall fill the child-space in my life with these kinds of companions. I smiled as my imagination fast-forwarded to a future self, nestled on the sofa with two perfectly house-trained old hens; in the background a veritable menagerie where canaries tweeted and parrots chattered, while outside in the garden a flock of geese and ducks waddled about. Maybe I could adopt a disabled raven or two.

Having assumed that I had zero experience of chickens other than the oven-ready variety, my new enthusiasm began to unearth memories to the contrary. The most potent was also the earliest. I must have been two or three when my grandparents left their large house in the Home Counties. I had thought I was too young to recall that place, but now I could summon to mind shadows of an orchard with gnarled and spreading trees, in the far corner a coop. My feelings were anxiety mingled with excitement – a dark, chaotic mass in the shade of the Pippins – how huge each hen had seemed, like something prehistoric, vicious and screeching. I remembered how gently my grandfather had treated them, even as he herded them away from me, his voice crooning 'all right, ladies, all right, my beauties.'

When I phoned my mother and asked her about this memory, she said yes, her father had always kept Rhode Island Reds. Even during the war when he was running a horticultural centre in East Africa, Grandpa had somehow managed to acquire some and kept them in amongst the avocados and acacia trees. In his retirement, he had expanded his collection to include ducks and geese.

Another memory was from only ten years previously. Pre-motherdom, I had lived alone in a top-storey flat in South London. It was when a relationship broke up that I decided the answer to loneliness was a pet. There wasn't a whole lot of choice – dogs: too much work; cats: no way for them to come and go; rodents: nasty, dirty creatures … I did my research, and discovered that what I truly desired in life was a bird. A sulphur-headed, red-cheeked one with a jaunty crest and elegant long tail – a cockatiel. I planned on its roaming free in the flat, learning to whistle the odd tune and even indulging me with a bit of chick-chat. I went out and started getting to know other bird-lovers; I bought a beautiful Art Nouveau cage to hang in the window ... Then I met Jay, and decided that a bloke was better company than a bird.

Since then, from bloke to babies, my life had taken a decidedly birdless path. Except that suddenly Roxy and Loxy offered the opportunity to backtrack. Jay could have been appalled, but he wasn't, he was intrigued. Especially if it meant freshly laid eggs and the wife did the mucky stuff. That was fine by me – I had ambitions; I was out to prove that my cockatiel plan had been a serious one – that birds can make excellent pets.

The larger of our two new hens was Loxy. With a body all round and red, she looked just like the Little Red Hen in the children's picture book. We knew the story well – how a feisty hen scatters her corn, then harvests and grinds it to make bread. She has to do it all alone because her animal friends are too lazy to help. But not our toddler Beatrice. Determined to demonstrate how good she was at helping, she would happily have spent all day, every day, scattering seed for Loxy. Perhaps 'scatter' is not strictly the term one should use for Bea's rummaging in the bucket, scooping up whatever would stick in her fist and thrusting the contents clattering across the patio. Loxy didn't mind – the closer her grain was clumped, the quicker she could gobble it up. The only problem came when the bucket was empty and Beatrice had nothing to do – soon enough she would grab at the fan of Loxy's tail, crying 'my chick chick, my chick chick' as she fled.

Not that Loxy was averse to human company. If you were prepared to forego the tail-grabbing, she was only too happy to be your friend. An hour spent sitting on the garden bench, and Loxy was bound to join me, kindly snipping short the grass around my feet. If I had brought a handful of corn, she would sometimes dare to hop into my lap to eat it. Cuddling Loxy like this made me realise why many mothers, including me, use 'chicken' as a term of endearment for their children.

In many ways, Loxy's style was more feline than gallinaceous. On summer afternoons, I could always find her in the south-facing corner of the garden, snuggled into a patch of long grass, her wings outspread.

If Loxy was the cat in the family, then Roxy was the dog. The skinnier and noisier of the two, she didn't approve of lazing around and if she found her sister sunbathing she might give her a spiteful peck. She was definitely inclined to bully, but at the same time possessed a playful streak. For example, if for some reason I had to put the duo away before the end of the day, I always went to Loxy first because I knew she could be easily cornered. Once she was safely locked away, I was in for the long haul. Roxy saw any attempt to catch her as an opportunity for a game of tag.

Sometimes I would try to ambush her while she was foraging – in amongst the bamboo was a good place for that. But even when I managed to creep very close indeed, at the last minute she would flit under my arm and away. If I pursued her, ludicrously emulating her zig-zag path down our narrow garden, Roxy would chortle as if in mockery – 'you can't catch me, you can't catch me!'

Another of Roxy's amusements was to strut up and down outside the back door. At first I thought it was because she wanted to come in, but when I went to turn the handle she skittered away. Eventually I realised she was watching her own reflection in the glass – attacking it and seducing it, taunting and testing it,

'Mirror, mirror on the wall.'

Eventually I opened the door and the spell was broken.

Chickens round the kitchen door were part of an ideal family life I had once dreamt of. One Jay and I had even tried. When Ellie had been six months old, we moved from our noisy London flat to a cottage in the Cotswolds. For a couple of years we pretended that life in the countryside was all about the wildlife, the space, our close-knit community tending the land. We tried to ignore the quad bikes and the SUVs, the fact that we and most of our neighbours were commuting away each day. In the end we returned to the city, reconciled to the fact that a truly rural upbringing was something we might never be able to offer our children.

But at least they could have a chickeny one. Chickens and children seemed ideally suited to one another; the daily routine of caring for them perfectly compatible. Though by nature a late riser, I had been disciplined by my daughters to get up soon after sunrise, even in summer. There was nothing for it, however thick the lining on their bedroom curtains, the day began with the light and one or other of them shouting the house awake. Next stop – Roxy and Loxy, also fretting to be let out. I learnt to relish the cacophony of birdsong, the smell of the dew – surprising rewards for performing the first-chore-of-the-day.

Their bedtimes were similarly in synch, except that the children had a lot to learn from the hens. I was charmed to discover that Roxy and Loxy knew exactly when they were tired and needed to go to bed. As dusk began to fall, the two of them would cease their business and potter down the alleyway at the side of the house. There they would huddle, patiently, until someone came to shut them in.

Needless to say, the bunny rabbit's natural schedule contrasted unfavourably with the hens'. Ellie was disappointed to discover that when she wanted to play, Snowy was dopey; when she needed to sleep, he was tearing around his cage looking for company. This unfortunate truth persuaded me that an essential quality in truly domesticated animals is that they keep the same hours as their owner.

They should also provide entertainment. Sitting and watching our chooks was excellent downtime for little Bea and me. We giggled at their funny way of walking, heads jutting forward and back, as if on a spring, their feet as big as clowns'. If the weather was dry, around midday they could always be found in the corner by my peonies, taking it in turns to enjoy their dust bath. As Loxy rolled around, she expressed her pleasure in appropriately feline terms: by purring.

And then she could growl. Deep in the throat this sound sat, her beak kept closed. It was definitely an early warning signal – telling that bitch Roxy she was not coming out of the bath until she was finished.

Unfazed by this, Roxy had plenty more impressive effects in her own repertoire. First and foremost, a noise disarmingly similar to a cock's crow. At first I worried it was a sign that she was one of those weirdos that lay a couple of eggs before transmogrifying into a male. People think this happens because they get an infection that renders their reproductive system useless; rather than having to admit their barrenness they pretend they have had a sex change. It must be this kind of avian sexual politics that inspired the old saying, *'a nagging wife and a crowing hen do no good for God nor men.'*

Fortunately for me, my bird's crow remained a cry of triumph to let everyone know she was far from barren. One long, high note using the head voice followed by six short, low ones in the chest signalled that a brand new egg had been born. My daughters loved to copy it; with some sexual politics-inspired lyrics from me, it became –

'Bra – not, not, not, not, not, not; Bra – not, not, not, not, not, not!'

Research from Macquarie University in Australia has established that domestic chickens have 20 different calls, falling broadly into two groups – food calls and alarm calls (the ultimate alarm call, naturally, being the one you get from a cock at 5am, though Roxy's egg-declaration comes a close second).

Food calls are interesting – members of the *Gallus* genus are said to have the richest array of food-calling behaviours in any animal group. Broody hens will use one particular series of clucks to draw attention to a tasty morsel; her chicks come running in response. When a cock is around, he will use exactly the same call to get his hens to take the earthworm he has found. He will also use it as a courtship song, or duping her into a quick shag, depending on how you look at it.

Roxy was always surprising me with her vocal range, and the apparent meaning behind it. She was partial to gargling. In its loudest, open-beaked version, it was less like water bubbling in the throat than stones rattling in a jar. I have a friend who swears her chickens can gargle the tune of Rossini's *William Tell Overture;* Roxy's version was rather more prosaic – signalling she was desperate to get out of her enclosure and have a run around.

She could also cackle with laughter, especially if she was excited by the prospect of a really good meal – '*What a bit of luck!*'. High-pitched shrieking she was very good at; almost as good as my children. When she and Loxy fought, they used the kind of resonance with which a soprano might break a wine glass. If they were very frightened, their scream was so shrill it could shatter a window pane. But generally they were contented to carry on their business with a perfectly conventional clucking.

Some researchers reckon clucking has developed as a way of keeping predators at bay – the constant stream of low-grade noise as the chickens flit about confuses that hawk hovering overhead. I would say the onomatopoeic suggestion in the word 'cluck' is less like the actual sound than a deep-throated '*bop*'.

'*Bop, bop, boppety bop,*' it has a subtly irregular beat. I think this is the sound Beatrix Potter intended to evoke with Sally Henny-penny's '*I go barefoot, barefoot, barefoot; I go barefoot, barefoot, barefoot.*'

I often thought Roxy and Loxy were clucking in order to keep in touch – even though they were each off in separate bits of the garden, finding their own tasty titbit. Other times they seemed to be chattering away independently, absent-mindedly, as I might have been doing were I still living alone in that one-bedroom flat. It could sound a bit grumpy but never suicidal. With their beaks closed it was similar to the squeak your fingers make when you rub a wet window. The open-beaked version was highly animated, full of tonal ups and downs, with hardly a break for breath. The underlying meaning was sure to be equally complex:

'So where are those lovely juicy daddy long legs? There was a nice crop of them yesterday, if I remember rightly. I dunno. What with this weather and all, I dunno ...'

Having been so delighted by my pets' vocalisations, I was disappointed to find a dearth of them in my classical music collection. Haydn's '*Symphony number 83*' earned itself the nickname '*La poule*' because of the fiddles' playful *appoggiatura*, contrasting with the oboe's dotted rhythm on one note. This combination is meant to sound very much like clucking and pecking; but I wasn't convinced. A composer friend who is quite obsessed with birdsong pointed to a hyperactive chicken in the '*Sonata Representativa*' by Biber; he even bothered to come round and play me Rameau's charmingly quirky '*La poule*' on the piano. But neither of them was quite chickeny enough for me; not after the real thing.

So what about chickens on stage? In his 1920s opera *The Cunning Little Vixen*, the Czech composer Janáček has a flock whose pecking gets the usual mechanical string treatment; they also sing. Unfortunately the composer was so smitten by his heroine, the vixen, that he gave her all the good tunes. The hens she kills are depicted as workaholic masochists, so numbingly institutionalised that their vocal lines lack any hint of bravura.

Janáček had a point. By the time he was writing his opera, most chickens round his way lived on large-scale poultry farms, soon to become the mass-production units of today. They had indeed become horribly institutionalised.

And yet in the very same period, just as chickens' voices were disappearing from most people's daily lives, musicians had started waking up to them. While in Czechoslovakia Janáček turned to farms, American jazz musicians went straight to the dirt and mess of a backyard much like mine. Big-arsed and busy, scrabbling about in the dust at the door, a flock of feathered females was ideally suited to Slim Gaillard's subversive art. Other musicians of the period, most notably Cab Calloway and Louis Jordan, were also inspired to write hilariously cheeky chicken songs.

From the 60s onwards the Funky Chicken came into its own. Where classical musicians had heard only regular monotony, the likes of Charles Mingus, Rufus Thomas, bands like The Meters, The Eliminators and the Chili Peppers ... all were transported by the syncopation of sound and movement. James Brown's saxophonist, Pee Wee Ellis, created a funk number called 'The Chicken' and it passed down to 80s bass player Jaco Pastorius who made it his signature tune. With a jerky four-time beat, shrieks and stabs from brass and woodwind, and a skittering about on electric guitar strings, its love of *Gallus gallus domesticus* is clear.

Back in my humble garden, it never ceased to amaze me how good Roxy and Loxy were at communicating. Not with the garden birds – they didn't seem in the slightest bit interested in them; their interest was in what we humans were up to. Who was that moving back and forth in the kitchen? Was that *Teletubbies* on TV again? Was Jay about to come out into the garden? And if so, did he have anything edible to offer?

Whenever they sensed we had visitors, they rushed up to the house to check them out. Sometimes, if we left the back door open we would discover them bustling about in the kitchen like a couple of bossy housewives, come to see that everything is up to their standards. When they wriggled their bouffant tails and deposited a Walnut Whip bang in the middle of the floor, Jay was appalled, but I didn't particularly mind. I shovelled it up and flung it straight out on the flowerbed. Then I put on my strict-Mummy voice, shooing the culprits towards the door, in the hope that gradually they might learn more civilised ways.

Not that I had entirely succumbed to my fantasy of house-trained hens. I recognised that they were happiest outdoors and we humans indoors. Catch me on one of my least sentimental days, I will tell you that chickens are not very bright, that they might be picturesque and charmingly friendly and all that, but mainly their talents lie in one, specialist area – eggs. In terms of food production, keeping egg-producers in the garden makes a lot of sense.

I am not an economist; I would say I am a thrifty householder and hen-holder, but I don't count every penny. That's why in the following calculations I am going to round my figures to the nearest easy number.

Roxy and Loxy were £10 each; their feed cost around £40 a year; their house cost nothing; add on a tenner for worming treatments and suchlike and you have spent £70 on the two of them. They each laid about 330 eggs per annum; that's 55 dozen in total. If you bought that many free-range eggs in the shops, they would cost you £150 – more than twice as much as you have spent.

Even though I maintain that keeping hens is cheap, things do become complicated if you start adding on things like your hourly rate for the labour. If you are a banker, for example, then I suppose even five minutes a day can make your hens look really un-economic. But if you are an unwaged mother like me, then you are seriously quids in. As food prices soar and City salaries plummet, which real economists reckon they might, then henkeeping could even start to look cost-effective for bankers.

Last but not least, along with all the pet-and-egg satisfaction, my chooks have introduced me to a whole new social scene. I had no idea quite how many people kept them or had kept them in the past – old and young; male and female; rich and poor; it is one of the few activities in this country that still manages to transcend social barriers. I find conversations about chickens are great ice-breakers, whether out shopping, at a party or at the school gate. Finding that your companion is one of the chickeny crowd brings a conspiratorial smile to both your faces – a chance to share one another's delight in bustling, chattering egg-producers. And once you're talking, you start discovering all the things you really should have known already, but that somehow, so far, had passed you by.

Cock 'n' Roll

He was like a cock who thought the sun had risen to hear him crow.

George Eliot, *Adam Bede*

Gradually it dawned on me that I had been too impulsive. I might, at the very least, have taken the precaution of scouring the net before I went chicken shopping. That way, I could have armed myself with all sorts of important information before Roxy and Loxy entered my garden. I might have discovered the existence of alektorophobia, and then I could have researched my rights and responsibilities regarding alektorophobic neighbours. Seriously. A thoughtful and considerate citizen never embarks on a major pet project without first finding out the rules.

Better late than never, I got on the phone to the council. Lara from the animal health department sounded friendly. She didn't think I needed to concern myself with my neighbours' phobias, as long as the garden was secure. She made sure I owned my property – tenancy agreements can contain prohibitive small print. She made sure it wasn't ex-council – their freehold stipulations can sometimes be anti-chicken. She told me if I was keeping fewer than 50 hens then I didn't need to get a licence. If the coop was of modest dimensions (meaning smaller than a large summer house) then I didn't need planning permission. If it was mobile, all the better.

'Can I keep a cock?' I enquired brightly, trying to sound as though the whole thing was still hypothetical.

'As far as we're concerned, you can.'

'What if the neighbours object to the noise?'

'A cock crowing is not an offence against any particular Act of Parliament, but there may be local bylaws. They would have to get in touch with the town council if they wanted to file a complaint.'

'Which department?' I asked, extra-specially mindful that forewarned was forearmed.

'Environmental health.'

'Is there anything else I need to know?'

'I can't think of anything. Most people just get on with it,' she chortled.

I couldn't believe it. Was this the same local authority that obliged me to have a new Criminal Records Bureau investigation every time I volunteered for an activity with my children – Sunday school, helping out with the school garden or a weekend's community camping? And every year insisted I renew them? This year for two daughters I had so far clocked up three CRBs, at £64 a shot. But for a flock of hens I needed nothing at all?

'I live in the centre of town,' I said. My greatest preoccupation since the Warrens' arrival had been whether I was providing them with enough space to qualify as 'free range'. 'My garden isn't big – only about 5 metres by 25 ... '

Lara patiently opened up her files and read out the rules on hen husbandry. According to EU agricultural policy (annex II, point c), I could keep 2500 per hectare. She got out her calculator ... that made 1 hen per 4 square metres. My garden had room for 31 and a quarter free-range hens. Crikey – what would the neighbours say to that?

A few days later, Lara phoned again. I wondered if somehow the neighbours had got wind of my plans and environmental health was already filing their complaints.

'Just to let you know,' she said, 'that we are getting a lot of enquiries at the moment ...' Surely she was about to say 'so we've decided to licence everyone after all ... £64 per hen'.

'So I thought I should take your address down, and put you on our database. Then if there is an outbreak of bird flu or something, we can come round and show you what to do.'

My conversation with Lara had turned out to be quite a relief. Rather than doing something bad through ignorance, I was doing even better than good. From her calculations, it seemed I was already providing my chickens with more than enough room. Even housing them on a balcony would still have been within the rules. And what about that cock idea? I had spotted some beautiful fellows at the sanctuary, cheaper than the hens.

Even as I was purchasing Roxy and Loxy, it had occurred to me that they might require a male presence in order to lay eggs. When the sanctuary lady failed to mention it, I decided there must be some kind of additive in their feed that provided the necessary hormonal stimulus. But now I knew the council didn't mind me getting a real, live, testosterone package; surely that would be preferable. I could buy one for a fiver and then give Lara another quick call, just to make sure of the bylaws.

In the back of my mind, I recalled Mick Jagger telling me I needed a little red rooster in order for there to be peace in the farmyard. It made sense. Without a cock, Roxy and Loxy had spent a lot of time and effort scrapping. They were fine when out and about in the garden, but I had been shocked at the way they fought over food, or snapped at one another's tails.

God forbid that a human 'hen party' has something to do with this kind of all-female behaviour – the most significant characteristic of such a group being the absence of that one, longed-for and assertive male. In avian hen parties, one female (in our case, Roxy) eventually fights her way to the top. The bigger the flock and the tougher the breed, the more this alpha female will assert her leadership qualities. Some use the exact same tactics their male counterpart might have used – crowing, mating with the others and even growing small spurs on her heels. But unlike the sex-change types, they retain their reproductive abilities. The human equivalent that springs to mind is Margaret Thatcher, 80s shoulder pads and all.

Were I to plant a male in their midst, Roxy and Loxy would cease their scrapping. Roxy the leader would instantly kowtow. And then, in proud assertion of his status, the master would start crowing, quite possibly very loudly indeed.

Like his wild cousins, the male blackbirds and blue tits, a cock needs to make his regular challenge to the world – checking whether a territorial competitor has expired, or a predator is approaching, he is probably letting everyone know that he is bold and brave and prepared to protect his unborn offspring at any cost.

He is also bang on time. A cock's announcing of the dawn has iconic status in our culture, from that weather vane atop a medieval church to the Kellogg's corn flakes packet. The Talmud refers frequently to the virtues of a cock that crows neither too early nor too late: the idea is that we humans must emulate his alertness, that our consciousness should be equally wakeful. He's there in all the Gospels, most famously in the Passion story as told by Matthew (Matthew ch.26, vs.34) when Christ predicts that his old follower Peter will betray him three times before the cock crows. The irony is that although Peter regards himself as utterly alert and faithful, following Jesus to the high priest's palace and attending his night-time trial, it is only when he hears the crowing that he wakes up to the fact that he has indeed betrayed him; but it's too late to go back.

These days, according to the poultry websites you can quash a cock's crow by housing him overnight somewhere light-proof and soundproof: indoors in a dark box or outdoors in a well insulated shed.

Not that Jay and I fancied a cock in the kitchen, and we had no shed. A sleeping compartment with a low ceiling might do the trick – he wouldn't be able to crow without stretching up his handsome head. But whenever he was out and about, he was bound to need to show off his vocal abilities. Over time, it might be possible to damp down the noise with judicious planting of bamboo or conifers. Or else we might just have to live with the neighbours' dirty looks and the odd run-in with environmental health.

A friend of mine who wishes to remain nameless went through endless battles because of complaints about his cocks' calls. The council threatened to fine him thousands of pounds; they must have had those bylaws in place that Lara warned me about; they also harked on a veritable smorgasbord of legislation that these days stands for good citizenship; he threatened to retaliate under European laws that stipulate certain decibel levels his cocks could not possibly have attained. Fortunately, the local Tory MP turned out to be fond of farmyard noises and the case was dropped.

On this basis, I wondered if we might get more than one. Contrary to popular belief, this was perfectly possible, as long as they were the right breed. Partridge Pekin bantams might have equated well to those mild-mannered fellows Mrs Thatcher controlled in her cabinet. My friend Sally has three of them; for the record, she also has neighbours with cocks, which meant that when they crow, no one can pinpoint the culprit. She didn't intend to have so many, but someone asked her to hatch some eggs and once the chicks arrived she let her emotions get the better of her and failed to cull the males. Out into the garden they went; her grandchildren christening them The Three Teds. One of them always takes the lead and the others follow behind as submissive as can be (clearly there is a pecking order between males too). Glorious fellows they are, The Three Teds, all decked out in gold and blue, strutting about the garden, upstaging the herbaceous borders with their vibrancy.

In purely aesthetic terms, there is nothing more appealing than a cock to upgrade your garden. Forget that reproduction Venus de Milo, or even a kitschy gnome. Apart from their disappointing lack of mobility, those artefacts tend to become submerged in living, growing things like plants. In contrast, a cock always makes sure he is noticed.

Of course, his cockiness is well celebrated in our language. You can cock your hat or your gun; you can be half-cocked, cock-a-hoop or cocksure; you can cock a snook, or simply cock up. There are cocktails and cockpits and cockneys; there are ballcocks and stopcocks, weathercocks and peacocks (previously known as peak-cocks). If the word does not originate in the cock itself (for example, the 'cockpit' being the small enclosure where they fight; or 'cockney' being a pejorative meaning 'cock's egg'), it is sure to indicate an erectile nature.

And it is not just in English that the word for a male chicken also refers to the *membrum virile*. In old Hebrew, *gever* refers to the bird, to the phallus and even to 'man' himself.

Ironically, avian cocks may look erect on the outside, but between their legs they have a hole (called the 'vent'), just like hens. It was Annie who taught me this. She's ten.

When Annie and her mother (my friend Katharine) went to buy a couple of Warrens and some Pekin pullets, the breeder said they should take the male bantam as well, free of charge. To Katharine it sounded a bit like the hen market's version of 'three for the price of two'. She tried to resist, but the salesman knew how to tug at both her purse and heartstrings – with his companions gone, he said, the cockerel would be picked on by the rest of the flock. They would be bound to peck him, or maybe worse.

At home, Annie's dad said they should wring his neck and make soup of him, but no one could bring themselves to do it. Such a handsome fellow he was, petrol black with feathery feet and a proudly arching tail. So they named him Brendan and let him out in the garden with the others.

It turned out that Brendan was quite a gentleman – Annie was impressed by the way he called to the hens when he unearthed a tasty slug for them; the way he always let them go first when the corn arrived. Most impressive was the way he defended them from next-door's cocker spaniel when it broke into the garden one day.

Annie said it was horrible to watch – little Brendan fluffing himself up to try and look bigger than he really was; flapping his wings and squawking while the hens hid under a bush. She grabbed at the dog, but he had no collar and escaped from under her hands, lunging at the cock. She screamed and ran inside to warn her mother, but Katharine just told her to be quiet. She didn't believe her.

When Annie came outside again, all she could see was a cloud of soot in the sky – Brendan's feathers, whooshing about. Then she saw that the dog had caught him in his mouth where he lay stiff and motionless. She screamed again, and luckily that startled the spaniel, who dropped his prey and scarpered. It was only when Brendan scuttled into the bushes with the others that she knew he was alive. His tail feathers had all been plucked out, leaving his bum bald as the parson's nose. But at least he was alive.

Annie told me she loved having Brendan for a pet. She found the sound of his crowing 'really beautiful'. When he 'trod' the hens, he always managed to trap them behind the hen house so they had a bit of privacy, she said. She had observed this activity quite carefully and decided that the hens didn't really mind: 'They don't scream – they just get on with it.'

After hearing Annie's stories, I decided the main reason to acquire a cock was for my children's sake. Brendan had furnished her with knowledge otherwise hard to come by when you are only ten. I could imagine our cock serving Ellie and Bea in the same way, and saving me a load of hassle when it came to sex education. When Annie told me that 'cocks don't have willies, they have vents' I was mighty impressed. She blushed only slightly when I asked for more detail – 'Dad told me that when he is shaking her around, his vent meets hers … the stuff … the sperm goes into her … into her egg … it slides in … and that's how she gets pregnant!'

It was my multiple-cock-owning friend Sally who disburdened me of the idea that hens need a male presence in order to lay eggs. They don't even need something extra in their feed. They lay perfectly well on their own, thank you very much. In fact, they probably lay better for not being hassled by a randy inseminator.

So there, Mick Jagger and your little red rooster, plus the whole cultural legacy that preceded you. It's not just the weathervanes and the corn flakes packets, Western culture has been confusing people like me right from the start. There he is in Aristophanes' satirical comedy *The Birds*, so grand that he is apparently King of Persia before the reign of Darius. In Plato's *Phaedo*, he flies in at the high-point of the drama where the great philosopher Socrates is committing suicide by hemlock poisoning:

'Crito, we owe a cock to Asclepius; please pay it; don't let it pass.'

There has been much discussion as to why such a wise man should have left the world with this particular statement. Scholars say his cock represents not simply material currency – that it embodies all sorts of moral and existential aspects of life. Personally, I am more inclined to regard the story as evidence that even great philosophers forget to prepare a good exit line.

Tactful as could be, Sally advised that it might be best to drop my plans for getting a cock. She warned that The Three Teds might be beautiful, but in general the male of the species do not make very good pets, sometimes becoming very aggressive, especially during the spring and summer breeding season. No fun at all, especially for small children.

That did it. No way was I having some overrated male icon flapping and clawing at my kids. I decided that, at least at this preliminary stage in my chicken-keeping career, a couple of hens would do.

And so to my cock-less flock. I had purchased two Warrens for my back garden, and from the council's point of view there was no problem in my having done so. Great. But I really did know very little about them. All I had thought when I bought them was that they would make excellent (egg-laying) companions for the bunny. I suppose I associated their name with his home. Or else I reckoned Warren was some antique family name … They looked just like hens in nursery stories, for goodness' sake.

It turns out that Warrens are the commonest commercial strain of laying hen, developed during the 1950s specifically for one thing – battery farming. These days their market name may also be ISA Brown or ISA Warren or Goldline. They must compete with all sorts of other battery birds with attractively bucolic titles – the Meadowsweet Ranger, the Bluebelle and the Speckledy. Goldline and the Heritage Skyline sound ever so posh, and there is real glamour in the Star varieties – Sussex Star, White Star, Black Star, Speckled Star and Amber Star. They are the result of laboratory-based cross-breeding and commonly known as commercial hybrids. There are currently estimated to be 40 billion of them in existence.

I say 'laboratory-based' because they derive from a very particular breeding process, meticulously controlled by scientists. Left to their own devices, chickens (or rather, cocks) tend not to be very discriminating in their choice of sexual partner. Most wild birds are choosy but these *Gallus* types, roaming free in jungles and villages, seem always to have been randomly promiscuous.

This proclivity came to the fore recently when Darwin's convictions about the origins of the domestic chicken were brought into question. Because chicken bones don't preserve well, no one can be certain of the period when she first got clucking. The best the paleontologists can come up with is that it was millions of years ago, somewhere in the tropical forests of South East and South Asia. Nevertheless, Darwin felt pretty certain that her one and only ancestor was the Red Junglefowl (*Gallus gallus)*: a beauteous tropical pheasant still to be spotted hiding under trees in Malaysia and Indonesia.

In 2008, Swedish biologist Leif Andersson and his team posed the question – how come that ancestor over there has white legs when our domestic birds have yellow ones?

After extensive research and whatever ancient evidence they could uncover, Andersson's team concluded that an early *Gallus gallus* must have hooked up with a yellow-skinned bird in order for the domesticated one to evolve. The most likely culprit was the Grey Junglefowl *(Gallus sonneratii)*: a lesser beauty with groovy yellow stockings, surviving well in many regions of India.

So – right from the start it seems, *Gallus gallus domesticus* was a hybrid, but not a commercial one. Here's an example of the difference: ISA Browns are often referred to as 'sex-linked'. This is to do with their strictly controlled origins, where different genes from their parents are used to determine gender-specific plumage colour. The ISA Brown hybridisation is said to have begun with two old traditional breeds – Rhode Island Red and Rhode Island White. The former possesses a 'gold' gene and the latter a 'silver' one. If you cross a Red mum with a White dad, all the chicks come out pale yellow. However, if you do it the other way around – dark dad and light mum, the boy babies are born yellow because of Mummy's silver gene, but the girls are buff coloured because of Daddy's gold one. This characteristic would be irrelevant in the wild, but is mighty convenient on a mass-production line, because the moment the chicks hatch we can keep the useful girls and chuck the useless boys in the grinding machine. Colour-coded chicks – what could be simpler.

The 'sex-linked' colour of their down is one of the few things we know about breeds like the ISA Brown. Huge commercial hatcheries produce tens of millions of them every year, but as to their exact genealogy – how much they contain of this breed or that, who their parents were and from whence they came … such information is never divulged. You can understand why: there is loads of money at stake. Imagine – if a hatchery creates a bird that can produce not 330, but 350 eggs a year, then all the poultry farms will be after her. A breeding cock, capable of fathering hundreds of thousands of mega-layers, will be kept under high security his whole life long.

Over the decades this lucrative and secretive business has become monopolised by a handful of multinationals. The 'ISA' bit of the ISA Brown, for example, comes from a French company founded in the 70s, merged with another big company called Hubbard in the 90s, merged again in the twenty-first century until it now has operational centres on four continents and produces 50 per cent of the world's hybrid layer population. Meanwhile, smaller operators have been priced out of the market.

Now, perhaps this would matter not one jot were the big corporations to follow the course of nature and keep expanding the gene pool. However, that's not how it works in business. To your average shareholder, a good chicken is not one that adapts to change, or ranges widely, or displays interesting quirks of nature. A good chicken is simply a machine that produces the maximum number of eggs with the minimum amount of fuss. Her qualities come down to things like sex-linking and 'efficient food-conversion' – in other words, not one penny wasted. She must have a docile enough temperament that she doesn't murder her fellows in the cage. In order for her to be fully focused on egg laying, she will have lost much of her mothering instinct. Using an increasingly narrow selection of birds, the hatcheries have gradually winnowed off the genetic codes that determine such characteristics, leaving the rest behind.

At the same time Leif Andersson was checking on the ancient genetic legacy, American geneticist Hans Cheng was doing the same for the contemporary one. For the US Department of Agriculture, he looked at a variety of genes in a few thousand hens. What Cheng discovered was that on average industrial hens lacked at least 50 per cent and sometimes 90 per cent of the diversity of their non-commercial cousins like the wild Junglefowl or the old Silkie breed. The genes they had left might be perfect for the demands of the battery farm, but there was nothing at all in reserve. Where were the rarities to help them survive when a nasty new virus came along? What about that mutating bacterial infection, those unforeseen changes to their diet or environment? What did commercial hybrids have on standby with which to adapt?

According to Cheng, chickens around the world are at increasing risk of something called 'disease shock'. Their lack of genetic diversity, fatally combined with their en masse living conditions, means a new sickness could wipe them out overnight.

I wish I had known this earlier. Had I done my research, I might have said no to Warrens and gone for a truly traditional breed whose precious gene pool needs supporting. There were probably quite a few of them hidden away at the sanctuary where I bought Roxy and Loxy, if only I had known to ask.

The American Rhode Island Red is the old breed my grandfather kept and most people have heard of. From the Netherlands came the Welsummer and the Barnevelder; from France the Maran; Leghorns arrived in the UK from the Italian port of Leghorn, Minorcas from Minorca, Hamburghs from … England. Also originating in England is the Poland. Then there are ones that actually sound like Brits – Old English Game, the Orpington, the Sussex …

Of the 150 registered breeds available in the UK, most were developed through careful cross-breeding during the nineteenth century. They may be referred to as 'pure breeds' but, as we now know, there was never such a thing in chickendom. The reason they were given this title is that during Darwin's period, breeders began the process of genetic control, preventing chickens from randomly interbreeding any more than they had already. Through a rigorous registration process, and the dedicated work of small-scale poultry keepers across the UK, they managed to segregate and protect their fine chicken creations against any more hybridization (natural or laboratorial). Many of these preserved breeds now have their own club and some even have their own heritage-rich website. They certainly look very pretty in the pictures, and would surely have enhanced the garden more than my Little Red Hens.

But they would not have been so easy to care for. As a beginner, I don't think I could have coped with one of those supermodel breeds, prone to all sorts of serious disorders. Marek's disease, infectious bronchitis, Salmonella, coccidiosis, Gumboro disease and epidemic tremor – I wouldn't have known how to recognise symptoms for any of these.

Fortunately, with Roxy and Loxy I didn't need to – they had been vaccinated against them at birth by large-scale spray-misting and, later, in their drinking water. They may have lacked the genes to combat future disaster, but they had taken plenty of medication to withstand quite a raft of present-day nasties. They were also less neurotic than most of their pure-bred cousins, less broody and less likely to escape from the garden by flying. It has to be said, their breeders knew what they were doing when they set about designing a low-maintenance layer.

I wanted my chooks to lay eggs – not just the three or four a week offered by the pure-breeds, but twice that, if possible. Between them, Roxy and Loxy produced at least a dozen a week; in my family's attempts to cut down on environmentally and financially extravagant meat, they met half our needs. I was contented with that. I had imagined that two happily free-ranging hens was all my garden had room for. Even after learning from Lara that I could have multiplied my stock tenfold, I still thought it had been safest to start small-scale.

And safest to shop close to home. From the pure breed websites, it is clear that most specialist keepers live hundreds or even thousands of miles away from my home in the south of England.

What would I have done if I had needed some specialist advice once I got them home? Or needed to re-stock?

Come to mention it, some of these birds are gob-smackingly expensive.

Meanwhile, a foray in the Yellow Pages uncovers various local hybrid agents – the middlemen of the commercial industry. They buy vaccinated one-day-old chicks from one of the gigantic hatcheries, feed them up for a few months (probably keeping them indoors all the while) and then sell them on at 'point of lay' (POL), at between 16 and 22 weeks. Instead of buying my POLs for a tenner from the sanctuary, I could have got them from one of these agents at half the price.

So there it is. Had I sat down and done all the research in advance, I feel that the outcome would most likely have been the same. I would have launched my henkeeping career in exactly the same way – with a pair of cheap and cheerful Warren pullets. I feel bad about Hans Cheng's research – the fact that the breeders have caused such dangerously low diversity. I worry about the potential 'disease shock'. Like so many products of the industrial age, from pesticides to the combustion engine, commercial hybrids may in future prove more destructive than they have been beneficial to life on planet earth. But in the meantime, people like me choose them because they so conveniently meet our needs.

There Ain't Nobody Here But Us Chickens

...There ain't nobody here at all,
So quiet yourself and stop that fuss
There ain't nobody here but us
We chickens tryin' to sleep, and you butt in
And hobble, hobble, hobble, hobble,
It's a sin.
Louis Jordan

Whether hybrid or pure, a hen needs a house. Unlike her wilder friends and relations, she is unable to create a safe one of her own. Once she became *Gallus gallus domesticus*, she relinquished responsibility to her domesticators.

Bought from the local pet shop or even off the internet, a coop can easily set you back a few hundred quid. Then again, if you want to save that second mortgage for another occasion, it is possible to spend less. Roxy and Loxy's first home was a two-storey hutch which cost nothing because (as its name suggests) it was really the rabbit's

When I first got them home, they were terrified out of their wits after their journey behind me in the car, cramped inside dark and smelly bags. Had I bought them in the late afternoon, it might have been fine – they could have dozed through the whole transition from sanctuary to hutch. As it was, in through the front door I charged around lunchtime, my hen sacks worryingly motionless. Having peered into their depths and established from the pairs of blinking eyes that the contents had not died of fright, what should I have done? Kept them in the hot and noisy house until dark, with the children constantly pestering them, and their droppings piling up? I dumped them in the hutch.

Jay had positioned it in the alleyway down the side of our terraced house, up against a brick wall, easily spied from the kitchen window and heard from our bedroom above. By the next morning we were rather more familiar with our pets' routines than we had anticipated. Snowy the bunny had spent the night thumping his back leg in warning against potential predators, rushing up and down the ladder, in and out of all the rooms. At six in the morning, when at last he settled, the chickens started chattering.

I found Snowy upstairs in his sleeping compartment, as far away from the noise as he could get, and the hens billing and cooing together in the bottom of the hutch. It was like one of those ghastly sleepovers when the children have kept one another awake all night, and everyone is grumpy for days afterwards, especially the parents.

Poultry books advise that you leave your new hens shut away for 24 hours before introducing them to something as overwhelming as your garden. When eventually you open up, they sometimes don't want to come out. At least, not until hunger and thirst eventually compel them to take the plunge. In my case, it was quite the opposite. Gently tucking Roxy under one arm so she couldn't flap, I inserted my fingers between her legs to prevent her scrabbling, just as I had seen the sanctuary lady do. It felt easy, almost as if I might have learnt the technique at my grandfather's heels.

Once the two chooks were out on the lawn and merrily pecking, I left them alone to investigate their new terrain. Only that afternoon, when I went to return them to the hutch, did I wonder if I was quite so knowledgeable about hen handling after all.

As Grandpa had so memorably demonstrated, when it comes to herding chickens, it is no use imitating their rhythm – getting all uppity. You need to stay calm and authoritative and always move slowly. Across the lawn I launched myself, arms outspread, knees bent and feet rolling carefully from heel to toe, hoping I looked a bit like a t'ai chi master. Jay stood guard at the main escape route, garden broom aloft. Perhaps we would have achieved our end, had it not been for our children. Someone had told Ellie she should raise her arms up and down in a vertical plane (apparently it confuses a chicken's vision). This technique, applied with five-year-old enthusiasm and mirrored with sound effects by her little sister, terrified Roxy and Loxy out of their wits. Off they ran, shrieking and flapping – one of them eventually digging herself under the shed, the other sheltering beneath the prickliest, thickest bush.

The first few days the hens were out, it took the whole family a good ten minutes to corner them and carry them back to the hutch. We could have resorted to a big fishing net on the end of a rod, but I reckon that might have frightened them off for ever. Instead we took to scattering a little corn and, while they were scoffing, gently swooping down and catching them. Along with the scattering, I introduced my farmer's wife imitation – *'Here, chick, chick, chick, chick'* which meant that soon enough I was able to impress my friends by simply calling out of the back door and the hens would come running. The daily expectation of delicious seed meant it did not take long for them to learn that being handled was really quite fun.

But being cooped up with rabbit was not. Snowy and the bantams might have looked good together in the shop window, but frankly what did a nocturnal mammal and two diurnal birds have in common? The hens' sleep must surely be wrecked (as was ours) by the rabbit's sport. Eventually I decided they should be separated. On a visit to the local tip I managed to salvage a modest-sized cage (those were the days when health and safety had not yet stepped in to disallow such risk-taking). Plonked on top of the hutch, with a bowl of food and one of water, this became the hens' peaceful new sleeping quarters.

An alternative, were I the DIY type, would have been to imitate my friend Paul from Zimbabwe and make their home myself.

A couple of years ago, as the political situation in his home country worsened, Paul and his wife Sekai made plans to bring their daughters to live in the UK. The children had been staying with family in Zimbabwe since their parents fled eight years previously. During all that time, Paul and Sekai had been in a state of limbo, never knowing from one day to the next if the Home Office might make a decision about their asylum status. They lived in a rented house and had never furnished it nicely or tended the garden, just in case they suddenly had to leave.

Before their daughters Kitty and May were due, a friend offered Paul some hens for which she no longer had room. And he thought – Why not?

In his home village they build houses for chickens out of large stones. About halfway up on the inside they insert horizontal poles that make do as perches. At the top of the four walls they build a frame out of sticks to support a roof made from savannah grass. These African coops look just like mini thatched cottages, standing side by side with the man-sized ones.

For Paul's English version, an old tea chest took the place of stone walls and a piece of corrugated plastic instead of thatch. He stood the chest with the opening facing upwards and cut out parallel elongated triangles from two opposite sides and joined up the lower end of these triangles by removing a rectangular section from a third side. Then he fixed his corrugated plastic across the sloping sides so the rain could drain off. A square of wooden batons attached to the underside of the plastic meant it rested in place and could be lifted to collect eggs or clean the inside. To prevent foxes from nosing it open, the roof was weighed down with bricks.

Bricks also provided the solution for keeping the floor dry – because English ground is so damp, Paul stood his chest on four sturdy brick legs. He cut a little door at the bottom for the hens to come and go, and screwed on metal hinges. Back home, the door would be hinged using strips of rubber from old car tyres; Paul says these are very strong and will keep out all the killers – English foxes are nothing compared to jackals and wildcats and snakes.

In Zim villages, cocks and hens roam freely during the day. Maize is the villagers' staple diet and there are always cobs left over to give the hens; they also get the husks of sunflower seeds, squeezed for cooking oil, and plenty of wildlife from their foraging.

When Kitty and Mary arrived at their English home, the first thing they saw was the chickens, running to welcome them. Paul phoned his parents to say the children had arrived safely, and heard them laugh at the country-clucking noise in the background.

For children with no money for computers or outings, chickens provide some recompense – Kitty and May spend happy hours feeding them, herding them and searching in the undergrowth to find where they have hidden their eggs. Once the children discover that place, the birds always go and lay somewhere else, but it's still fun looking for the new hiding place. Just the other day, the hens found a little grass snake in a bush and tried to kill it. Kitty rescued the reptile and brought it inside, but the chickens weren't having that – they followed, strutting around the kitchen until she gave them back their snake.

Recycled cage or recycled tea chest, the ideal position for a chicken coop is not necessarily down an alleyway at the side of a terraced house. After a good ten months housing our hens there, we had a visit from the next door neighbour. He said they were keeping him awake.

'But they sleep when we do!' I protested.

'Only if you get up at five in the morning!' he retorted.

His proposal was that we send them out of earshot to the end of the garden. I didn't like it. A couple of dozen metres away from the house felt terribly far.

'I have nothing against keeping chickens,' my neighbour proffered. But I felt somehow he must. He probably thought they belonged in a faraway farmyard, not lovingly integrated into domestic life.

I said, 'What about the fox? He doesn't dare come right up to the house, but he's bound to visit them down there.'

'Most people consider it more hygienic to keep their livestock at a distance.'

He sounded like he'd done his research.

Eventually Jay persuaded me that moving Roxy and Loxy was not such a bad idea. And that keeping in with our neighbours was a very good one. In many ways, the rougher, messier nature of the bottom of the garden would suit them better than our patio. Ellie and Bea could think of them as our fairies down there. On a more practical note, the trees and shrubs would provide plenty of shade and wind protection. Being jungle natives, hens get a sense of security from having jungly stuff around them, and I reckoned they were unlikely to spot the difference between a buddleia and a palm.

Then I decided I might as well go the whole hog and splash out on a home designed for chickens rather than for rabbits.

In terms of housing design, chicken coops, like dog kennels, seem not to have advanced with the times. I don't know from which period Paul's stone cottages derive, but the traditional British coop looks decidedly Victorian. There is something just a bit too idyllic about it, reminding one how perfect country life once was compared to our nasty, industrialised cities. If expanded to a human scale, these folksy homes would be the perfect place for Red Riding Hood or Hansel and Gretel. These coops are made of timber which, as every fairytale character knows, needs a regular daubing with weather-treatment to keep it waterproof and prevent it from cracking and warping.

Or else you can forego that pleasure and spend your money on contemporary plastic. For Habitat fans, the eglu is an excellent lifestyle choice for hens. My friend Fiona has two. These smooth capsules, in a range of bright colours, look as though they just landed in her garden from outer space. If you are concerned about sustainability, the label says it is 'made from energy efficient polymers using modern construction techniques; the eglu will last for years and at the end of its life can be 100 per cent recycled.'

If you are really concerned about sustainability, then an even better bet is the Eco Hen House from award-winning Scottish company Solway Recycling. Built from 100 per cent recycled plastic (maybe it's an eglu that got thrown away), in design terms it sits somewhere between the modernist chic of the eglu and the earthiness of a traditional coop. It is quite a bit cheaper than its competitor.

Though not a fan of plastic, even recycled, I appreciated the fact that these contemporary creations were both chicken- and keeper-friendly, with good-sized doors and removable sections for easy cleaning. Fiona said the main advantage over traditional designs was their lack of nooks and crannies, those nailed-down felt-coverings and awkwardly pitched roofs where tiny chicken mite reside. A smooth surface means those mite can never get cosy.

But chickens can. Some people find plastic doesn't adapt to temperature change as well as wood, but Fiona considered her eglus versatile enough. On a really sunny summer's day she might put them in the shade of the trees, and on a horribly cold night she slung a blanket over the top, just in case.

Being a jungle creature, a hen's optimum environmental temperature tends to be that of an average British home – around 21 degrees Celsius. However, on a frosty night it is probably best not to bring her inside, as once she gets used to central heating it could be mighty troublesome getting her to re-acclimatise. Fiona's hens manage to keep their body temperature up during the winter by eating more during the day and huddling in their double-walled eglu at night. They get through the odd cold snap by chasing each other up and down the garden a couple of times a day. Apparently you can protect their combs and wattles from frostbite by daubing them with Vaseline, but Fiona has decided this is more relevant for Iceland than England. She trusts their plumage to do its job just as it would inside her duvet – fluffing up and trapping the body heat.

As long as it doesn't get wet: wet feathers mean chilly chicken. Which is why the coop roof is such an important feature to consider. To keep out the rain, it needs to be well-pitched and close-fitting, but not too close. Paul's square of corrugated plastic was perfect because it had gaps between the ridges to release stale hen-breath. Hens do a lot of breathing. Seriously – they need a lot more oxygen relative to their body-weight than we humans, which perhaps explains why the Victorians introduced good coop ventilation long before they got round to the same in hospitals; it wasn't just because they loved their hens so much. 'Good coop ventilation' means some sort of window, air hole or door, preferably high up, where the bird's breath sails away rather than condensing and creating the moisture that makes her chilly.

Good ventilation also prevents overheating. A hen's normal body temperature is high – around 41 degrees Celsius. Lucky her – it guards against infection. Towards the end of the nineteenth century, when Louis Pasteur was looking for ways to vaccinate against killers like rabies and cholera, he discovered it was this heat that gave chickens their immunity.

But during hot weather a hot body can be problematic. While a duck keeps cool by paddling in cold water, and a human perspires discreetly, a hen must open her beak and pant. Though her tongue is not soft and pink like a dog's, it still manages to do the job – the water vapour rising from its surface to cool her system. Inside a badly ventilated house it will condense, leaving the poor chook prematurely stewed.

Having considered the roof, you need to think about the door. The official term for a hen's front door is the pop-hole – according to the manuals, it should measure no more than an A4 piece of paper and be facing away from prevailing winds. It has either a sliding shutter or a drop-down ramp that opens and closes with the aid of an electric winch-system and a light-sensitive timer to tell it when to do so. One problem with such technology can be if it interprets glowering clouds as darkness and locks your hens outside just as a major storm is breaking. Then you have to go and buy a clock timer, and remember to alter the opening and closing times for British Summer Time.

Personally, I was not attracted to all this gadgetry. I rather enjoyed the old-fashioned ritual of shutting up my hens at night and letting them out first thing in the morning. My new coop required nothing fancier than what I had previously in the cage – a front door with a serviceable catch.

I did need to improve on the perching facilities. Though Roxy and Loxy had managed for a year without one, a rod of wood is pretty essential for happy hendom. Like all birds, a branch (however high) is where they feel safe from predators while they sleep. In conventional coops it is generally positioned at least ten centimetres from the floor and is four centimetres wide, slightly rounded at the sides for easy gripping. If you can find a branch of similar dimensions, hens apparently prefer it to the machine-made version – they like to exercise their toes around its nobbles and bends; their wild ancestry feels safest with something asymmetrical. Each bird needs about a span's length of rod or branch; give them much more and they will simply huddle together at one end. They do poo in their sleep, which is why the perch needs to be raised high enough for them not to be sitting all night in their own mess. It should also be easily removable for cleaning, with newspaper or a droppings tray underneath.

One essential my rabbit cage had not lacked was the sleeping compartment. Not that sleeping was what Roxy and Loxy used it for. It was where they deposited their eggs.

If you have ever taken children on a farm visit, you will have come across nesting boxes. They are the sticking-out bits at the back of the hen house where the children are encouraged to form an orderly queue. Have you seen the henkeepers' trick? Under the lid of the nesting box, little Jenny discovers a lovely warm egg; she lifts it out, proudly presents it to the keeper and turns away, satisfied her job is done. Just at that moment, and before little Jimmy shuffles to the front of the queue, the very same egg (freshly warmed by Jenny's hot little hand) slides back under the lid.

In essence, the nesting box is a convenient way for the henkeeper to collect his or her booty. If it is the darkest place the hens can find, then they feel compelled to lay there, foolishly assuming that darkness means privacy. Paul's tea chest did not have one, which is why his hens imitated their Junglefowl ancestors and laid their eggs under bushes.

The ideal nesting box should be neither too big nor too small – its dimensions are roughly what you need to store half a dozen bottles of wine. Like Goldilocks, a hen needs to feel just right in there. And like baby bear, she shouldn't mind sharing her facility, though some of the larger coops provide several boxes.

To stop a hen using her nesting box as a toilet, it's useful to have some sort of shutter to close it off. The floor needs to be lined with something to give the eggs a soft landing – shredded bank statements might suit the contemporary chic of the eglu. Of the more organic options, untreated wood shavings are the least mite- and mould-friendly, but straw gives it that real-nest look. Hay can contain fungal spores that give chickens respiratory problems.

As for the outdoor space, like Paul's chickens, mine used to run straight out of their sleeping quarters into the garden. However, if avian flu comes along and the Department for Environment, Food and Rural Affairs (DEFRA) gets its act together, this kind of free-ranging will become illegal. It is already so for people who mind about the lawn.

All hens, especially hybrids, get their kicks by scoffing grass and scratching up grubs – so much so that within a shockingly short amount of time Roxy and Loxy rendered our few square metres of grass post-Apocalyptic. Having made the decision to exile them to the bottom of the garden, the best damage limitation I could think of was to confine their foraging. Fortunately, that area was well-drained – boggy land is more easily infected with parasites and generally unpleasant underfoot.

The most conventional way to confine a flock is to fix a run to their coop. If DIY Paul had had more time, he might have knocked one up for me using timber and wire mesh. Jay did offer, but I decided instead to go internet shopping, and thereby discovered the amazing array available.

Some looked modest enough, with walls only a couple of feet high and a mesh roof so Mr Fox couldn't hop in and Roxy and

Loxy couldn't hop out. Some had no roof at all, which looked inviting to Foxy, unless you had Colditz-style fencing around your garden. Others were more of an aviary – totally enclosed and tall enough to walk around in. Some versions were interlocking, so you could expand or contract the space to suit your flock. Some came with a coop that had two pop-holes; you took it in turns to attach the run to whichever side you were using.

This last design is where the concept of rotation comes in. The idea is that after a certain amount of time being scratched and pecked and pooed-on, the limited area inside the run is wrecked. If by some miracle the grass has survived, it may well be singed by the burning effect of nitrates from the droppings; it possibly also harbours nasty parasites and infectious diseases. In winter, this patch could do with a break even though the grass won't grow back – an opportunity to give it a sprinkling of lime to help reduce the acidity of the soil created by the droppings. Hens should be kept away from the limed area for a good month afterwards, as lime is bad for their tummies. Strictly speaking, for the ground to fully recover, it needs a good few months' rest, which means rotating the run to four separate areas. Many people make do with two.

My problem was that even with a two-patch rotation, most of these coop-run-combos were too big to fit in the bottom of my garden. As every estate agent knows, in these crowded British Isles every metre counts. So how was I going to give Roxy and Loxy room to roam?

In his marvellous 1920s tome, *Fowls and How to Keep Them*, Rosslyn Mannering describes Londoners 'successfully managing and rearing' chickens on their roofs. One of my ambitions was to find a creative carpenter to build a coop-run combo for my friend

Esther who lives in a top-storey flat off the Marylebone Road. I was sure Esther's balcony was big enough for some sort of multi-storey structure where a couple of happy hens could run around and scratch, have perches and nest boxes and all the things they love. If land-dwelling *Homo sapiens* can live in vertical formation, then how much more easily should *Gallus gallus domesticus* manage it, with that urban jungle heritage.

In the meantime, the most high-rise accommodation I could come up with was on a mere two storeys – sleeping and nesting quarters above and a run below. Apart from its obvious space-saving features, a major asset was the way the upper room protected the run beneath. Neither rain nor frost nor blazing sun was going to hinder my hens' daytime activities. It was also the kind of cover my friend Lara and her bosses at DEFRA would be insisting on if avian flu came along.

Having inspected numerous two-storey arrangements, I eventually decided on a sturdy triangular structure called an Ark. At over £400, it was one of the most expensive on the market but after almost a year of henkeeping I felt confident I was in it for the long run; the time had come to invest in infrastructure. My Ark came with excellent credentials from a company with 30 years' experience, and the word 'Ark' reassuringly in their title – *Forsham Cottage Arks*.

The Boughton, as mine was named, looked like a mini mountain hut – rather than Red Riding Hood, it would make a perfect home for Heidi. Its steeply sloping sides were originally intended not for snow but to prevent sheep climbing up. A pair of smooth handles sticking out either end meant it could be carried between two adults, or dragged by one when the hens needed 'rotating'. The walls were mesh on the run downstairs turning to solid slats of red pine upstairs. One of the walls was fitted with handles so I could lift out the whole side to clean. A nesting box hidden under the eaves had a ventilation hole at the top and its own little door that slotted in and out – perfect for childish hands in search of eggs. At either end, the bottom section of wall slotted in and out and was held in place with a couple of revolving catches.

Many people recommend replacing simple catches like this with bolts and padlocks. If you don't believe me, take a look at the Poultry Websites and you will find numerous missives from devastated correspondents who pottered out one morning as usual, only to find someone had nicked their flock.

Living in a terraced house, I did not feel too vulnerable to the chicken thieves. Nevertheless, if my hens had been costly pure-breeds I think I would have invested in locks. The birds exhibited at poultry shows are apparently particularly vulnerable because thieves are on the prowl there, getting details of where to come and steal them; unless such birds are microchipped, they are quite easily sold on. If I had owned a charismatic gamefowl I might have gone for the whole security shebang – wireless infrared alarms or even CCTV. Though illegal for more than 150 years, cockfighting is still popular in the UK and the owners of the most vicious fighters make good money.

I think the name 'Ark' appealed to me on security grounds. I rather fancied it as an Ark of the Covenant, containing something precious and awe-inspiring. And like Noah's, it would surely protect its contents, if not from the flood then at least from the fox.

My modern-day Noah arrived from Worcestershire with his walls flat-packed in the back of the van. By the time his creation was erected, it was clearly never going to get back through the house again. Ah well. If ever Jay and I sold up, an extra hen-home at the end of the garden might prove an asset. It certainly looked good – like the beginnings of a miniature Alpine village, its pine walls melding with the surrounding tree trunks.

That evening, as Roxy and Loxy were nodding off in their cage, one by one I carried them gingerly to their new quarters. I wasn't sure how I was going to get them on to the perch if, after so many months in the cage, they had lost the skill. For all I knew, they might never have had it, as I hadn't noticed any perches at the sanctuary. With the side-wall of the Ark removed, I eased them on to their rod of wood and watched fascinated as their perching reflex curled each claw tight and each ruffled body found its balance.

Between the two storeys of their new home hinged a solid ladder that levered up and down from the outside by means of a cord fixed over a hook on the roof. I raised it to the closed position to keep out night-time draughts and perhaps even a predator or two. In the morning I went out to unhook it and watched my duo scramble downstairs to their daytime quarters. This seemed absolutely natural to them, just as it was natural to take themselves upstairs to bed at night – like all the other birds and even me.

In my opinion, whoever designed the Chicken Ark really worked hard to combine its inhabitant's needs with her keeper's. Accessible and mobile for me, it provided all that Roxy and Loxy could wish for, including a sheltered play-area.

By the end of the very first day, they had dug themselves a sand-pit and were playing at stuffing their bodies with dirt. Under their wings and between their feathers it went; once they could feel it close to their skin, they would scrabble themselves deeper and deeper into the trough formed by their bodies, sometimes rolling right on to their backs, legs absurdly akimbo. I loved to watch them curling and stretching, drawing the earth up under their arms and letting it fall away again like so many waves of imaginary water.

Historians tell us that when the Emperor Napoleon was on campaign and had nowhere to wash, a valet used a 'flesh-brush' to clean his bare skin. It is to this same practical end that a hen takes a dust bath. After much rolling and flailing, the grit falls away, carrying with it all the bits and pieces that have been itching her – especially mites and lice.

Of course, a mud bath is not the same thing at all. Which is why the British climate is not brilliant for hens, who would far prefer the Med if only they knew. Though the Ark run was sheltered, still the rain managed to angle itself to enter through the mesh, and seep in from surrounding land. When things got really soggy, I threw in some wood chips – they soaked up most of the mud. In autumn I raked up fallen leaves and bagged them for the same purpose. I know now that my original ruse of chucking in bark chips that would otherwise have mulched my allotment was not a good idea – they can carry a dangerous fungus.

In defiance of my nation's weather, I set about creating my own dust bath. Into an old plastic seed tray I shovelled a mixture of topsoil and silver sand (not builder's, the chemicals therein cause all sorts of bother), which was the grittiest combination I could come up with. This I placed against the solid end-wall of the run, to shelter it from the prevailing wind (and rain). If ever it got wet, I chucked the contents on the flowerbeds and refilled it. In winter, when we had wood fires indoors, I added cinders swept from my hearth. Chicken bliss.

The one and only EU stipulation with which I could not comply was the one given me by Lara from the council regarding space. To qualify as free-range hens, Roxy and Loxy were meant to have 8 square metres of run (4 per bird). Mine measured only half that. And they did look cramped in there, side-stepping along the mesh walls from morning till night.

My solution? After all that fuss about the lawn, I gave in and let them run free in the garden during the day. But this time I blocked off the bits I could not longer bear to see bald.

Chicken Feed

We got ground to dig
And worms to scratch
It takes a lot of sitting
Getting chicks to hatch
Louis Jordan

Having splashed out on my Ark, it seemed only right and proper to acquire complimentary feeding kit. For a whole year Roxy and Loxy had made do with entirely inadequate facilities – an old bowl for drink, another for food; grain and water constantly spilling and mixing to a gruel under their feet. The next step on from this would be to acquire 'Grub and Glug' pots from the makers of the trendy eglu – a pair of hollow orbs, moulded together like some disembodied bosom. My friend Fiona recommended them, but I decided I needed a system where feeder and drinker were entirely separate. Attached to the wall of the run was a hopper for the hens' feed. To keep the water clean, I fancied a classic half-gallon water fountain.

I could have bought a cheap one made of white plastic that you fill (somewhat awkwardly) upside down, but it would soon have worn out and didn't have much aesthetic appeal. For my money, the one made of solid galvanised steel set much the highest standard in industrial chic. Shaped like a hurricane lantern, this prop would look great swinging from my hand on my daily trips down the garden.

It was also a clever piece of design technology. Near the bottom of the main cylindrical can was a little hole where water could trickle out into the surrounding trough. Over it slid an outer cylinder (with the handle at the top) that revolved a few degrees into slots to hold it in place. As the water rose high enough to cover the little hole, it created a vacuum, thereby halting the outward flow.

According to Rosslyn Mannering's *Fowls and How to Keep Them*, it is possible to make a simple home-made version of this on the same principles using a bottle, inverted and hung in a metal bracket over a shallow pan. Had I not been on a spending spree, I might have tried it. Instead, I decided that the half-gallon galvanised water fountain was an excellent investment that would last a lifetime (mine, not theirs).

The day it arrived, Roxy and Loxy somehow managed to kick dirt and poo into the trough, and continued to do so on a daily basis. My solution was to raise their drinker out of the firing line, on to a couple of bricks. Nevertheless, I found that every time I refilled it I needed to scrub out both inner and outer cylinder. For this purpose I posted an old washing-up brush by the outdoor tap, and on occasions even took the bottle of Fairy Liquid with me to knock back the algae.

Just like humans, hens thrive on clean water. If it is frozen over on a cold winter morn, I add boiling water from the kettle and watch it steam. For something even fancier, I sometimes add a tablespoon of cider vinegar to keep the poultry in tip-top condition. For a while I added a special tonic that gave it a pretty pink glow (the water, not the poultry) – a teaspoon-full for every couple of litres. The label on my bottle said the ingredients were 'water, sugar and minerals'. I was interested in its high iron content (1,000 mg per kg, apparently); it also contained phosphorus, manganese and copper – Cheers!

Watch carefully while a hen is drinking, and you will see how supple is her neck, slinking in and out as she scoops up water. Because her tongue is less malleable than ours, she cannot easily direct a mouthful of liquid towards her gullet. She needs the help of gravity. Back her head tosses, beak wide, allowing the water to slip down her open, upstretched throat like a Cossack quaffing vodka.

When it comes to her eating style, the similarity is with a less assertive sort of human being: as the proverb tells us – she is toothless. Or, to be more precise – her one and only tooth is for piercing her shell at birth; it falls off her beak after about a week.

According to the paleontologists, until about 80 million years ago hens' teeth were not so scarce, and more recently someone succeeded in reactivating the dormant tooth gene, stimulating some unhatched chicks to grow a full set of ivories. Unfortunately, all these little mutants died before they hatched, so for the next few million years the proverb will still hold. And the hen will continue to make do – firstly with a sharp beak that is extremely adept at slicing off whatever morsel of slug or grass she fancies; secondly, with her little pointy tongue that tilts that morsel down her slippery gullet.

And then what happens? One important characteristic a chicken does not share with a human is a stomach. Instead, she has a two-part mechanism: a storage sack called a crop, and a primitive food mixer called a gizzard. As she forages, the un-chewed food gathers in the crop positioned down her front where a bib might otherwise hang. Pick up a hen that has just eaten corn or pellets, and you will be able to feel the ball of grains she has accumulated there. Slowly (normally when she is asleep), the food passes from the crop into a glandular stomach where digestive acids and enzymes moisten it and start to break it down before continuing on to the gizzard. This horny bag the size of a golf ball functions with the help of a load of grit that she has swallowed on her rounds. Its muscles expand and contract, jiggling the grit around and thereby mashing her feed as her ancient teeth might once have done.

And if it's not slug or grass she is sending down there, then like as not it's something that her keeper supplied in her hopper. Chicken feed. In everyday English, this word may suggest something inconsequential and trifling, but for the dedicated chickenkeeper, that couldn't be further from the truth.

By the time Roxy and Loxy came to live with us, just like every hybrid POL before them, they had been weaned off their baby food and on to what is termed 'compound feed'. Produced in a factory and containing all the nutrients required by a hybrid layer, this is the most convenient form of chicken feed, used by farmers and small-holders alike. Basically, it is a man-made mega-fuel.

Compound feed comes in two forms – 'layers' pellets' and 'layers' mash'. The difference between pellets and mash is much the same as that between cubed or granulated sugar – the former has been squeezed by a machine into bite-sized chunks. Because of the work that has gone into shaping them, pellets are slightly more expensive, but considered by poultry farmers to be more convenient because the nutrition is at its most concentrated. Some people recommend mash for birds confined to a run because it takes them longer to eat, giving them less time to get bored, poor things. It needs to be stirred before placing in the hopper because the calcium content may have dropped to the bottom.

My first 20kg of pellets cost less than a tenner and came with Roxy and Loxy from the sanctuary. The lady there advised me not to bother weighing the feed when I gave it to them – just to top up the feeder whenever it became empty. I had bought it with an accompanying sack of corn, but when I scattered the mixture for the hens, they always picked out the tasty seeds and left the weird-looking pellets behind. If I could have been bothered, I would have given them pellets in the morning and corn as a treat in the afternoon. In the end I gave up the corn option – after all, 'compound feed' meant the pellets were designed to serve all their nutritional needs. I thought I might just take a look at the label and see what that entailed.

The first ingredients were *Lutein* and *Zeaxanthin* from marigold extracts and *Citranaxanthin* from citrus fruits. Talk about 'compound feed'! It's not the hen that benefits from these, but rather the person who eats her eggs. All three are versions of *Xanthophyll* – the chemical compound that gives yolks their yellow colour (and which free-range hens pick up from grass and other greens). In fact, they provide no nutritional value for either creator or eater, but they do make breakfast more aesthetically pleasing.

Then came the main ingredients – the filling carbs: barley, maize and wheat. Then 'non-GM soya' for much-needed protein, a 'vegetarian vitamin mineral pre-mix' (meaning the vitamins didn't have a gelatine coating), plus *dicalcium phosphate* (for general wellbeing, especially bones) and limestone (for good, strong egg shells).

The label also reassured me that my pellets contained 'no hexane-extracted ingredients'. Apparently, the solvent hexane is used to extract the goodness from the soya bean meal and may be carcinogenic. I think probably consumers are worrying about hexane getting into their own food rather than their pets', but I could be wrong.

For the hens' sake, a 'best before' label warned me when the pellets might start to give them a tummy ache. This tended to be around three months from the manufacturing date (and, with any luck, the date when I bought my sack), which was another reason not to supplement with corn as it took all that time for my two to use up one sack. I could have bought smaller bags, kindly filled by my local pet shop owner: that's if I wanted to pay twice the price. Instead, I tried to be organised about buying a new 20kg sack only when the previous one was finished and always checking the 'best before' label.

I stored my pellets outside in a steel bin. My plastic bin with clip fasteners was probably more airtight but a ravenous mouse or rat could have gnawed its way through the bottom, so instead I used that one for bedding. There they stood, my two bins, side by side outside the kitchen window, like a pair of sentinels keeping guard over my Good Life.

Except that when you look into it, soya-based feed is not such a good thing at all.

The first problem is that it has to be imported. Unfortunately, Great Britain does not have a suitable climate for growing soya, not even in poly-tunnels. The closest place is Italy, but the Italians can't produce nearly enough to fulfil our needs. UK companies that use soya, not just in processed animal feeds but in human ones too, have to go to faraway places like Canada or China and in the process give their products a socking great carbon footprint.

But it's worse than that. Over the past couple of decades, Europeans have got very worked up about the genetic modification of soya seed. Meantime, the Americans get along nicely with their fully modified version, exporting enough of it to satisfy 46 per cent of the world soya market (and, incidentally, raising the seed price extortionately). People who are passionately against GM may well be eating US soya unknowingly, hidden away in burgers and biscuits and breakfast cereals. But they are also driving an expansion in demand for non-GM soya by 10 per cent a year.

And how can this demand be met? If people like me want to pay less than a tenner for a sack of non-GM soya feed, we need it grown somewhere cheap, somewhere warm and somewhere big enough to continue expanding production. Where better than that vast tract of virgin land, the Amazon rainforest?

I have been there – a couple of weeks living in a hammock on a steamboat; never had I seen a place more teaming with life, more throbbing with fecundity. That rainforest is home to nearly a tenth of the world's mammal population and a staggering 15 per cent of the world's known land-based plant species. It is also home to the Amazonians themselves: my boat was their means of transport; a trip to market meant several days on board, often with all the family. Dressed in jeans and T-shirts, they brought with them sacks of firewood or fruit or turtle meat, returning home with computers and televisions. By day they hung about in the heat, playing cards and gossiping, often three or four to a hammock. By night they drank home-made *cachaça* and taught me to dance lambada on the roof of the boat. They were cheerful but life was hard in the face of the usual problems: poverty and HIV.

Then came the soya boom. Faced with some feed merchant's contract, my steamboat companions must have leapt at the opportunity. At last they would be able to afford medication; an education for their children. All they had to do was slash and burn the forest to make way for the crop. No one said anything about the longer-term effects of polluting the waterways and eroding the soil.

Just to add insult to injury, these soya plantations turn out to have a major impact on global warming. Forget ocean algae or oak woods or any other form of carbon capture – the Amazon rainforest absorbs carbon dioxide faster than anything. It is the carbon sink of the world. In destroying it, we are exacerbating the problem of climate change in a really big way.

Which is why that innocent bin of mine was not really a sign of the Good Life at all.

When I started telling my chicken-keeping friends about the bad, bad soya industry they were concerned. A few were appalled. But all of them told me there was no alternative.

'You don't want to mess with a hybrid's diet,' they said. 'She needs really good protein to produce all those eggs.'

I said, 'But couldn't we find some other source of protein?'

They said, 'Don't you remember BSE?!'

Before BSE, most of the protein in industrial animal feed (including layers' pellets) came from meat by-products like bone and fat. It took all those spongy brains to teach us that it is not such a great idea for animals to eat one another.

'But have you ever heard of mad-hen disease?' I asked.

'We need to look after the food chain,' said my friends, gravely.

'What about looking after the rainforest?'

'If you think about it,' said one, 'a daily egg is such a miraculous feat that it is bound to require some sort of sacrifice!'

Maybe I needed to try a different tack.

'This is industrially processed food,' said I. 'The amount of grief I get keeping my daughters away from that stuff because it's bad for the environment, bad for them, and not even as tasty as the food I can cook myself!'

No one could disagree with that. When you looked at those pellets, so rigidly uniform, so dusty and unappetising, you didn't even think McDonald's; you thought supplements. They were the sort of things astronauts took when stranded in outer space, not what any sane being would choose to eat in the comfort of her own home. For Roxy and Loxy, the equivalent of outer space might have been the anonymity of their destined battery farm, but they hadn't gone there; they had ended up living with me, in my home.

And I had ended up eating their eggs. When you taste something salmony in a shop-bought egg, it is probably because of something its mother has been eating. And that isn't salmon; it's soya. According to the latest research, there is a problem in the way some hens metabolise seeds like rape and soya – it seems to cause their intestinal microbes to form fishy-tastes in the eggs. Yuck.

My friends conceded that I had a strong case. They agreed that my home was not a production line; it was one place where I could positively eschew the evils of industrial agriculture. They agreed that Roxy and Loxy did not particularly like eating pellets, and that salmon-flavoured eggs were disgusting. They were unconvinced I would come up with an alternative. But I wasn't.

It didn't take much surfing to discover that, in fact, people in the agricultural industry were already trying to do something about our dependence on soya. In the US, the law allowed farmers to feed their poultry on pellets made from 'spent hens'. That did not feel like a solution for me, but an interesting challenge for the BSE/food-chain folk.

A bit more surfing and I discovered that tests were underway in Europe to try and establish how locally grown beans can provide sufficient protein for high-performers like Roxy and Loxy. One of the problems the researchers have come up against concerns herbicides and pesticides: where soya requires minimal use (that's GM for you), peas and beans need loads. On the plus side, their production is less energy-intensive than soya-bean meal, partly because the seed doesn't have to travel halfway round the world. Most importantly, we don't have to destroy our carbon sink for them.

Even closer to home, I found that the organic grocer Abel & Cole was successfully fattening up its 'rainforest-friendly chicken' on a mixture of peas and beans, wheat, rape and sunflower seed. I reckoned a small-scale urban henkeeper might take a tropical leaf out of their book.

Next stop – the feed merchants. The same company that supplied my pellets would happily sell me sacks of 'super mixed corn'. A kind of muesli, it contained whole grains like barley and wheat as well as peas, though its label did warn that it would NOT provide a 'nutritionally balanced diet'. What this meant was that my hens needed to supplement their breakfast cereal with extra amino acids. Additional forms of protein.

Of course, before either industrial animal-product or soya came along, a major source of protein for hens was the world around them. It was just as well Roxy and Loxy still had access to a large chunk of my garden – I loved to watch them as they ventured forth from their run, snatching at leaves, flitting their little heads all the while to keep an eye out for predators, just as their jungle ancestors must have done.

Most of their protein-rich food came from scratching away the surface of my flowerbeds. Scratching was a whole technique in itself – how each bird would lower her centre of gravity slightly in order to stay stable, her head still up and alert. Out those big feet shot behind her, one by one – sweep, sweep, she shifted the ground aside, exposing fresh earth beneath. Only then did she risk a look downwards – quick as a flash, that squirming grub was in her beak and her head was up again. Sweep, sweep; off she went again.

Each time they were out in the garden and caught a juicy worm or a centipede, their amino acid count must have soared. In summertime, they were picking up all sorts of bugs as well – daddy long legs were particularly popular. Foraging like this was excellent exercise, and probably my chickens' favourite form of entertainment. It was infinitely preferable to those pellets.

And there were ways to improve the foraging potential in the garden. For starters, Jay and I should leave the lawn un-mown; even bounded as it was by chicken wire, long grass was the perfect breeding place for bugs. The roses would be left to ramble, as would the clematis and hop and jasmine. When I pruned the buddleia or the fruit trees, instead of packing them laboriously into the council waste bag, I would henceforward leave branches and foliage stacked against the fence so that tasty beasts might gather there. It was a very attractive prospect knowing that the less I tended things, the more bountiful would they prove.

I also realised that behind every flowerpot and under every slab hid a delicious meal or two. I found gathering up slugs and snails and feeding them to the hens much more satisfying than scattering poison. I even endeared myself to my neighbours by inviting them to lob theirs over the fence. The children found great entertainment in herding the chickens towards the latest crop of molluscs and watching them tweezer them from their shells, or simply thrash them against the patio.

The other protein-packed delicacy on the menu was frogs. If the chickens were rummaging around near the pond and I heard a noise like a rubber bath toy being squeezed, I knew they had found one and were starting to torture it to death. I did try to intervene – I like frogs.

But hang on a moment. Was 125 square metres of city garden really enough space in which to forage all the protein they needed? If they were NOT getting that 'nutritionally balanced diet', how would I know? Might they even stop laying?

Back at my computer, I searched out the Eden Farm research centre not far from my home and put my question to their hen specialist. Her immediate response was that the likelihood of Roxy and Loxy stopping laying was very small. Hybrids are genetically predisposed to produce more than 300 eggs a year, she reminded me; if they are getting a poor diet, then it shows up not in a reduction in number but in the condition of the eggs. After a moment of deliberation she confirmed –

'I think all a lower-protein diet might cause is smaller eggs.'

'That's fine by me,' I said, 'my family rather likes them small'.

But then I had the idea of protein-rich milk products. Hens just love buttermilk, as any Victorian fattening up their Sunday roast will tell you. What I could offer was the remains of the children's breakfast cereal, doused in cow's milk. Plus the dregs of each bottle, washed around with a little water. This I poured over whatever leftovers were hanging around in the kitchen.

Ah, leftovers. Some people feel that they are not acceptable food for chickens. I have heard it said that hens 'are not rubbish bins'; that giving them scraps is somehow undignified and dirty. One book I read said no more than one fifth of your flock's diet should be made up of scraps, though weighing them to find out struck me as unnecessarily arduous.

I have always been keen on leftovers. Born in the 60s to a family not fully recovered from wartime frugality, I fell in happily with childhood diktats about not wasting food. As Great Britain entered a new era, where the average household threw away at least a third of its food, I was stubbornly reheating yesterday's pasta.

And then my children came along. Most mealtimes young Beatrice would change her mind several times about what brand of cereal she liked, whether she wanted pesto on her spaghetti after all, whether potatoes were actually quite disgusting … Argh! Loxy and Roxy arrived just in time to save me from a tsunami of leftovers. As I scraped the food from my daughter's plate, a huge consolation was that at least my chooks would gobble it up. I even used this fact as an effective taunt – any hint of 'I don't like it,' being met with 'Great, the chickens will!' and miraculously the meal became acceptable after all.

But an invitation to other people's meals could be testing. I cringed at the sight of all that precious food being tipped into the bin; meanwhile Bea grinned – at last Mummy was facing the fact that she was seriously out of touch with the zeitgeist. I stood my ground: how could people donate £500 a year to the starving nations and yet chuck away food worth triple that? How could Bea's nursery school be making all that effort to raise funds for Save the Children when the slop bucket from dinnertime could feed a whole refugee camp?

Slop bucket. If you have ever read *Charlotte's Web*, you will recall an intelligent pig's point of view on slops: a veritable smorgasbord of pleasures. With BSE the bottom dropped out of the slops market, leaving Chinese takeaways drowning in oceans of wasted pig feed. Poultry farmers say they never really liked it because it does what it says on the label – it slops about. It also rots a lot quicker than dry feed.

But in the comfort of your own home, why not go sloppy? I would like to start a campaign to bring back slops. I would dearly like to keep a pig and feed it on all the food Bea's school throws out. My lesser achievement (so far) has been to give some to the hens. I took to presenting the teacher with an empty bucket twice a week, and the dinner ladies filled it for me with whatever scraps remained on the children's plates.

The quantity of food I carried home was far more than we could use. When I thought about it, the frugal bit of me was angered by the sheer profligacy of this generation who thought meals were for throwing away. The more tender-hearted bit thought such an excessive volume of leftovers might be my fault: early on in my campaign I had given the children a talk about chickens' favourite foods being exactly the same as theirs (Cheerios, raisins, cherry tomatoes, that sort of thing). I printed out some photos of Roxy and Loxy eating these treats and the teacher kindly posted them on the wall where everyone could enjoy them. Young Ben was so inspired he even took to bringing in what was left of his supper in a little plastic bag – noodles, rice, pizza ... Perhaps it was generosity of spirit that motivated these children to leave their meals on their plates.

The teacher was ever so keen on the ritual; she typed up and laminated a list of foodstuffs that should stay OUT of the bucket – banana peel; rotten or dirty food; citrus (especially orange peel); and meat. She ruled that salty food was a bad idea for both chickens and children, so were whole raw potatoes, and so was too much acid all at once (like cooking-apple peelings).

As for the hens – they were in ecstasy. Normally, I would say they were dainty eaters, but served with nursery leftovers it was a different matter altogether. Suddenly each one transmogrified from gentle herbivore to vicious carnivore. Grabbing a chunk of food in her beak, she drew it away from the plate, blocking out that other vulture with her body so there was less chance of competition. Rip, thrash, gulp. Then back she came for more.

Only once she had finished off all the choicest delicacies did she wander away to wipe clean the blade of her beak. First the right side and then the left – Swipe, swipe, against a branch or a rock. Swipe, swipe again. Having indulged in such filthy feasting, she was meticulous about returning to her pristine form.

*

I must admit I did not keep to the school rules about leftovers. The no-meat rule was to show that the teacher and I were well aware of fears about BSE. However, having discovered that American hens were officially allowed to eat one another, at home I did let the odd bit of chicken meat fall into my slops bucket. Sausage-ends and lasagne often found their way there too. Though I did avoid feeding the hens whole raw potatoes, one of their favourite things was potato peelings zapped poison-free and soft with a few minutes in the microwave.

The most abundant foodstuff in the school bucket was carrot, chopped into unnaturally uniform orange sticks. I was keen to offer it to the hens in large amounts, aware that carrot is regarded as a natural way to purge their system of worms. Unfortunately, the sticks proved as inedible to Roxy and Loxy as they had done to the children.

They were much keener on discarded sandwiches. Some people say you should not feed chickens bread because it gives them a tummy ache, yet we all know from childhood how their duck cousins love it. I decided that mouldy bread was a bad idea, but a modest amount of stale sandwiches or cake soaked in milk was a delicacy I couldn't deny my birds. In fact, according to the Roman, Cicero, if fed on soft cake, a hen can be effectively consulted as an oracle (*de Divinatione – Concerning Divination,* Book Two). When you need a prediction, you open up her cage and feed her the cake. If she stays in her cage, makes noises, beats her wings or flies away, the omen is bad; if she eats greedily, it is good.

Of course, in this day and age we are more aware of the dangers of a carbohydrate-heavy diet. Frankly, no hen (even one offering a good omen) should stuff herself too full with it, or else she will miss out on more nutritious alternatives. I think this means offering things like bread and cake often enough that she never feels the need to be greedy. I once stayed with a peasant farmer on the island of Ischia who cooked up a huge vat of pasta every morning for himself, his dogs, his cats and his hens. After such a hearty all-Italian breakfast, the hens spent the rest of the day rooting out slugs and snails and other goodies from the surrounding fields. They were in noticeably excellent health, and their eggs were scrumptious.

I didn't keep pasta in my kitchen cupboard especially for my hens, but I did get a stash of porridge oats. When I first acquired Roxy and Loxy it was wintertime and a hen-loving friend recommended cheering them up by giving them a hot meal each morning. I remembered how my grandfather used to go to quite a bit of bother, putting yesterday's kitchen scraps in a saucepan and boiling them to a stinking gruel. Me, at breakfast time I did my twenty-first century thing and whacked some oats, water and a little milk in the microwave, along with whatever tasty morsels were hanging around from the night before. Often I delivered the dish so piping hot that my diners had to throw it all around to cool it down.

From a chef's point of view, hens are easy to cook for because, unlike children, they have very few tastebuds. Where the average child will have around 10,000, a chicken has only a couple of dozen, situated at the base of her tongue and down her pharynx. It seems that nature has granted her only enough taste awareness to avoid the bitterest of plants that might poison her. There is variation from hen to hen – for example, some are happy to down the spiciest of curries, while others would reject it. But generally speaking, potent foods can bypass her taste system with alarming ease.

It was intriguing what exactly Roxy and Loxy were getting out of their favourite foods. I suspected that the attraction of little treats like tomatoes and raisins was mainly in their size and texture. The oat porridge was warm and comforting in their gizzards. The poultry spice I added to their mixed corn for a treat when they were under the weather contained all sorts of stimulants – ginger, turmeric, aniseed and fenugreek. It cheered them up no end, probably because it improved their digestion. And though the olfactory sense in *Gallus gallus domesticus* is thought to be extremely limited, they must have been able to pick up its exotic aroma.

My only disappointment was that the spices failed to resurrect themselves in the hens' eggs. People had warned me that I would get interesting variations in egg flavour when I started feeding my chooks their exotic slops diet. At the weekend, we stood by for the fish pie from Friday's school dinner to make a second appearance at breakfast, but it never did. The same with garlic, at least as far as I was concerned. For several months, I was feeding the hens mashed garlic cloves because it was said to be good for their digestive systems. No one noticed. Only when Jay found out what I was doing did he detect something garlicky in his omelette and so ordered up a powdered version for the hens, guaranteed not to influence the taste of their eggs. Fortunately, an unanticipated side effect of feeding garlic powder to the hens was that it made their droppings less malodorous, for which our neighbours were most grateful.

We also ordered up a grit and oyster shell mixture, but Roxy and Loxy didn't seem to want it and it remained untouched in their hopper. Eventually a chicken-keeping friend said they were probably picking up quite enough grit for their gizzards as they ranged around the garden. Instead of oyster shell, she suggested that I could provide essential calcium in the form of their own eggshells, recycled.

I remembered my grandfather carefully baking his hens' eggshells in the oven, grinding them up with a pestle and mortar and feeding them back to his Rhode Island Reds. My more informal version of the same tradition was that when I used eggs in the kitchen, I dropped their shells on to a tray in the bottom of the oven where they happened to get baked whenever I cooked something in there. I just had to make sure I didn't let them have more than one sitting, otherwise it was carbonised-shell time.

Next on my menu were the all-important greens. It is not an old wives' tale – a dark yolk is a nutritionally superior egg, as long as the colour is the result of Mum's eating lots of greens. According to US research, the eggs of chickens roaming on pasture contain 34 per cent less cholesterol, 10 per cent less fat, 40 per cent more vitamin A, twice as much omega-6 fatty acid, and four times as much omega-3 fatty acid as eggs from industrially fed hens.

Once I got rid of those industrial pellets, I started to see the difference between a yolk coloured by marigold extract and one coloured by grass – the former has a psychedelic saffron glow, while the latter is sunshine yellow tending towards emerald. Now I could see whether they were getting enough greens; and much of the time, I am ashamed to say, I found they were not.

If I had had a half-acre field, I would have made sure I shifted my Ark across it daily, giving Roxy and Loxy the chance to gobble up large amounts of grass. As it was, at the bottom of my town garden I was shifting it roughly monthly, from one patch of bare earth to the other. In the meantime, on the hens' daily forays abroad, the lawn (though busily breeding insects) was out of bounds. The best greens on offer were geranium and honeysuckle, nasturtium and alchemilla. In winter there was little even of these.

The allotment proved far superior. An early ruse was to take an old rabbit run down there and whenever I went for a few hours' digging, I took one of the hens with me, hidden in a canvas bag out of sight of the local dogs. As long as I positioned the run in exactly the right place, it was entirely a win-win situation: while I dug, she got to peck up all those weeds I would otherwise be on my hands and knees pulling up. The only problem was that Roxy and Loxy soon began to regard this activity as not worth the unpleasant trip in the bottom of a bag. Before long, whenever they saw me approaching with it, they skedaddled.

They knew full well they wouldn't lose out; soon enough I would return bearing delicious offerings. Comfrey (especially the bocking 4 cultivar) was a particular favourite, as were the plants whose names suggest people have long known who likes eating them – chickweed and fat hen. If none of these could be found, it was always worth bringing home uprooted dandelions or chard that had gone to seed. During the summer, cow parsley was popular; in winter, cabbages.

I also boned up on which plants to avoid. In alphabetical order, but probably not comprehensive, they are: aconite, bracken, bryony, buttercup, daffodil, delphinium, dock (seed), foxglove, hellebore, hemlock, henbane, horseradish, horsetails, hyacinth, hydrangea, ivy, laburnum, larkspur, lily-of-the-valley, lupin, nightshade (including tomato and potato plants), oleander, privet, ragwort, rhododendron, rhubarb (leaves), St John's wort, yew.

Of course, it turned out that Roxy and Loxy had been surrounded by these plants in the garden, but had never touched them. Such was their survival instinct. But what if I had thrown old potato plants or rhubarb leaves into their run? They might have been bored enough to try a bite, if it wasn't already soiled and trampled underfoot.

In retrospect, it was not such a brilliant idea to throw any loose leaves from the allotment on to the floor of their run. It took only a matter of minutes for them to become entirely unappetising. My guru, Rosslyn Mannering, advises that we henkeepers string up our greens in a net, high enough that the hens have to jump to peck them. A great believer in exercise was Mannering. But that was the 1920s; after well-nigh a century's worth of genetic refinement, my chooks were no more likely to jump for their greens than my children were.

Their style was more snatch and grab. Once I cottoned on to this, my solution became to string up a garland of greens on the inside of the mesh of the run. Not so high that they needed to jump, but at beak-height and fixed tightly enough that they offered resistance. When Roxy and Loxy came to snatch off their morsel of leaf, the stems held fast.

I did the same thing with sunflowers. Since giving up soya, I reckoned that home-grown seeds might provide the chickens with just as high a level of energy. At last I had something to do with my thousands of pumpkin seeds – I took to drying them on the windowsill and adding them to the mixed grain. But sunflower was an even greater delicacy. From around October, November, just as the colder weather set in and the hens were getting hungrier, my plants had perfectly shrivelled heads, bursting with seeds. I cut each one with a good length of stalk and left it to desiccate in a basket by the back door: a bouquet of prickly black orbs, ready to be presented to Roxy and Loxy over the winter months. One by one I carefully wove each stalk through the mesh.

It was well worth the bother of growing and harvesting and weaving. Come the spring, my plump and healthy hens would return the favour with a glut of creamy, fresh eggs.

Eggs

It has, I believe, been often remarked that
a hen is only an egg's way of making another egg.
Samuel Butler, *Life and Habit*

From day one, eggs were a major focus in my henkeeping. Once installed in their cage near the house, there was little for Roxy and Loxy to do, thought I, but lay some. Every morning I would rush downstairs, fling open their door and rummage around in the wood shavings like a child at the lucky dip, sure that today my treasure would be there, hiding in the corner.

But nothing. A couple of days before Christmas, I was out carol singing with a group of friends and someone introduced me to a woman who kept chickens. She welcomed me to the sorority with such warmth that I decided to let on about my birds' fundamental shortcoming.

'Do you know how old they are?' she asked. I wasn't sure; had not thought to ask. 'They might not be ready, you know. Even if they're full grown, they could still be a month or two off laying … Some breeds take six months to get going.'

For all I knew (but did not say), I had been sold birds whose egg-laying days were already numbered.

'Perhaps the trauma of the move might have set them back,' I suggested, trying to stay positive.

'Maybe,' she agreed.

'Have you tried pot eggs?'

I had to admit I wasn't sure what she meant.

'You can buy them at Countrywide – eggs made of pottery. You put them in the nest and sometimes that encourages them to lay.'

'OK.'

'But of course your main problem is the season.'

'Of course.'

My new friend and I exchanged knowing smiles, as if both fully aware of the link between chickens and seasons.

Back home at my computer, it turned out that just like humans, chickens suffer from SAD (Seasonally Affective Disorder). They are super-sensitive to light and as the days get shorter they conserve their energy by reducing their egg-laying activities.

Commercial breeders have done everything they can to get rid of the SAD genes in hybrids, but they have not entirely succeeded. This is why poultry farmers use electric lighting through the winter months – it cheers the hens up enough to think it is worth their laying. If you want to try this trick at home, you can place your coop up against the house and leave the curtains open for the evening. The light from your living room may be enough to fool them. However, it is worth bearing in mind that each hen produces a finite number of eggs – either she strings these out over the year, with the help of your local power station, or she keeps her work seasonal.

I bought a couple of lovely white eggs made of clay, but it didn't seam to inspire Roxy and Loxy, so I gave them to the children for their cookery games and decided it was such an extraordinary thing to produce an egg at all that my chooks shouldn't be expected to lay out of season. It was right and proper that they should take a winter break. Little did I realise this break would last so long that I would forget the whole egg-laying plan entirely.

It was a fine morning in the middle of February, Shrove Tuesday to be precise, and I couldn't think why Roxy wasn't coming out of the sleeping compartment; she must be ill. I went to check – she looked in the rudest of health; when I stretched out a friendly hand she went at it with a peck. I retreated to finish the washing up, and there at the sink it started to occur to me that she might be broody. As I was starting to scroll through the poultry-breeding forum, she began to shout,

'Bra – not, not, not, not, not, not; Bra – not, not, not, not, not, not!'

At which point, without even going to retrieve the evidence from the wood shavings, I clicked straight out of nesting and over to recipes for pancakes.

Have you noticed the lack of 'small' chickens' eggs in the supermarket? As sure as eggs is … if ever you go searching for them you will find nothing between the incy wincy quail's and a great big 'medium' one. Large can look as though it came from a dinosaur. The only reason I can think for this is that, in keeping with packets of crisps and fizzy drinks and packs of butter, these days only the Behemoth will do. And because small eggs are no longer required in the supermarket, the commercial hybrid has been engineered not to produce them. Only when she is a young thing like Roxy, just starting out on her laying career, might she produce something of modest size. Which is how I remember that first egg – a golden colour with a smattering of speckles at the pointed end, small enough to nestle in my palm; it was so perfectly bijou it might have been fashioned by Fabergé. I cracked it in half and made celebratory pancakes.

A few days later, Loxy presented her offering: pinkish and even teenier, its contents were all white – no yolk at all. This kind of egg is called a 'wind' egg and is a sign that the pullet has not quite got her act together. We ate it all the same and rather hoped she would produce more. Alas – the next one was normal.

Of the two hens, Loxy was the slightly less reliable layer, preferring to wait until the afternoon to produce hers. Taking a leaf out of God's book, she would often give herself one day off in the week. In contrast, Roxy created one golden egg every morning without fail: her 24-hour cycle even more efficient than the most reliable battery hen's (who is expected to have one of 25.5).

In keeping with their differing productivity, they also had unique laying styles. To produce an egg, Roxy needed several hours' sitting in the sleeping compartment, while Loxy took far less time but required absolute peace and quiet. Any kind of disturbance (like her sister wanting to come in) and she would burst out, screaming her head off. Eventually she found herself a private spot underneath the acanthus bush where she could lay without disturbance. Until Roxy followed suit and laid her eggs in the same place. By the time I cottoned on, there was a good dozen sequestered there.

*

As soon as our hens got laying, egg collection became an essential part of the family routine. Each morning five-year-old Ellie would race outside to open their door, ignoring poor Snowy in the cage below. Not long after, she would appear in the kitchen with breakfast clutched between her palms. As official breakfast-maker in our house, Jay was the one who got to cook it, with little Bea in her high chair making sure she got her fair share. If Ellie brought only a single egg, Daddy must make a big thing of taking out the one Loxy had laid the day before and scrambling the two together so both sisters got absolutely equal shares of brand-new and nearly-new egg.

My great pride was that, unlike most of their peers, they were fully aware of where their breakfast came from. Meanwhile, I was starting to wonder about all our previous breakfasts: I really knew very little indeed about all those shop eggs I had bought and eaten over the decades.

I was aware that a code stamped on the egg identified its origins. I had also noticed a striking red lion indicating British Lion Quality. But I had never thought to investigate further.

Apparently, the lion indicates that the producer belongs to a national 'hygiene and traceability programme'. In the interests of food safety, this stamp gives a 'best before' date – 27 days from when it was laid. As for the code, working backwards: the long number at the end identifies the registered flock from whence it came, 'UK' refers to the country where the flock lives, and the lonesome number at the start dictates what sort of egg it is.

I had known to avoid the cheapest eggs in the supermarket, produced by miserable caged birds. I knew not to be attracted by words like 'farm fresh' on the box – the packing company had merely omitted the word 'battery' from in front. But what about 'barn eggs'? What exactly were they? Was there a dust bath in that barn and room to run around? And what exactly was the difference between an organic egg and a free-range one? I had already gathered from Lara in animal health that the official free-range space entailed a modest 4 square metres per bird – nothing like the acres of gently wooded pasture I had once imagined. But what about organic birds? Did they get more space, and even the odd daddy long legs?

'Organic' is signalled by a lonesome 'O'. The EU stipulates that these chickens require only the same space as free range which means they have no more chance of catching insects in summer. But they are likely to belong to a smaller flock, which means life is less chaotic and competitive. They eat food containing no laboratory-produced chemicals or ingredients that might harm the environment, but it doesn't mean they avoid soya-based feed or get enough greens to produce naturally yellow yolks.

Because of the high cost of farming organically, these eggs are the most expensive and according to DEFRA they attracted only 4 per cent of the market in 2009; that number is going down. In contrast, 'free range' is an increasingly popular choice. Providing 3,289 million eggs in 2009, it was 37 per cent of the market and that figure is rising fast (by nearly 20 per cent in a decade).

The number '1' indicates 'free range'. These flocks (however large) must be given continuous daytime access to the outdoors from the time they start laying, but whether they use their privilege is another matter. Hens are creatures of habit – having grown accustomed to communing around indoor feeders and drinkers, they may need to be encouraged towards an alternative. Another thing we henkeepers know is that their jungle nature means they are unlikely to want to go out if there is nowhere to shelter, no bush or tree.

Which is where the RSPCA's 'Freedom Food' scheme comes in. There it is on the box – a circular logo with an 'f' swinging across the middle (almost every single free-range egg producer in the UK is signed up to the label). This tells you that the farmer who produced these eggs was working with animal welfare advisors, constantly researching and reviewing practices. The provision of natural shelter such as woodland and hedgerows is exactly what they have been working on recently, so let's hope the hens will feel inspired to use their range.

Meanwhile barn hens (with a number '2' on the shell) don't get the chance. The RSPCA reckons barn hens don't have such a bad life, but people like me feel less sure, and their eggs therefore constitute only 4 per cent of the market (around 1 million hens). From the EU's point of view, the hens can belong to a gargantuan flock, but have space to run around, raised perches or platforms to sleep and some sort of dust bath. The barn floors are carpeted in 'deep litter', sometimes with mesh for droppings to fall through.

As for the number '3's – in 2009, there were 17 million battery hens in the UK and their produce formed around 55 per cent of the market. These eggs are not quite as simple to avoid as I had assumed. It was all very well my leaving the shelled eggs on the shelf, but what about all the unshelled ones? Take a peek in my fridge and you would find a large jar of mayonnaise without any mention of what kind of eggs went into it. In the larder sat custard powder and luxury egg pasta, both similarly minus the information. By default, this meant they were made with battery eggs.

Worse still, because egg products don't have strict labelling, I had no idea whether their ingredients came from countries with non-existent welfare standards. For all I knew, my custard powder was made with dried egg from some giant battery farm in India. And this problem is likely to get worse. With new EU legislation coming into force in 2012 to improve the welfare of hens, British eggs are likely to rise in price, thereby doubtless expanding demand for the cheaper imports, both shelled and unshelled.

Which all goes to show that things can never be as good as they are in your own backyard. Only there do you know exactly what food is going into your birds, exactly the amount of natural shelter and mud and dust they are getting from day to day. Without the use of a single stamp, our homegrown eggs were 100 per cent free range, fresh, happy, soya-free, Salmonella-free, E.coli-free, and very nearly free of charge.

Not that my children were aware of their breakfast having such superlative credentials. All they knew was that their pets had produced it, and this simple association pleased me immensely. To have fed them and stroked them and looked those birds in the eye seemed to me the essence of responsibility when it came to eating eggs.

What I wasn't prepared for was the rather more existential question that occurred because Ellie was so regularly burrowing under their bottoms.

'Mummy?'

'Yes, my sweet?'

'What exactly IS an egg?'

There we sat at 7.30 in the morning with our tea and toast and eggs, the news on the radio distracting me.

Mmm. Well. What a time to launch my sex education responsibilities.

Of the two cells that come together to make a baby, the egg and the sperm, the egg is the larger, less mobile one; all the growing and the feeding happens there. It is the same for snails; the same for Mummy mammals like me. The difference between my eggs and Roxy's is in what biologists call our 'reproductive effort' – mine is somewhere well under a millionth of my body weight; miniscule. Whereas each of hers is about 3 per cent. In a year of laying, Roxy will convert about nine times her own weight into eggs; a quarter of her total energy goes into the act.

'If it takes so much energy, why doesn't she have a rest?' asked Ellie.

Because you keep taking her eggs away! In the bird kingdom there are two styles of egg-laying – the 'determinate' layer and the 'indeterminate' layer. In the former group are the blackbirds and the sparrows and all those other garden birds we know so well. During the nesting season, they lay a set number of eggs at a time; if the magpies come along and steal some, they have to make do with a smaller family. In contrast, an 'indeterminate' layer like an Indian Junglefowl slowly accumulates eggs in her nest (in the Junglefowl's case, about 12 glossy brown ones, laid over two to three weeks); once that clutch is completed, if a predator removes one, she will happily lay another to replace it. Given as much food as she wants and a predatory human or two, she might go on for ever.

Which is exactly what Roxy and Loxy are doing. Of course, over the year a domestic hen will lay far more eggs than she could ever hatch. But she may remember to lay them in the same style as her jungle ancestors – in other words, having accumulated a clutch, she will take a break. That's what Loxy was doing when she stopped laying for a day – completing her clutch, before that five-year-old predator came along again …

Phew. It's time for piano practice and school and for me to get researching for next time. By then I shall have my *Encyclopedia of Poultry* on hand, plus Rosslyn Mannering and my other chicken guides, plus my food science manual … I'm ready.

'Mummy?'

'Yes, lovely.'

'How does Loxy make an egg?'

Well. When Loxy was born, just like you and me and most female animals, she had several thousand microscopic germ cells in the sack of her ovary. But her body did not grow symmetrically like yours and mine – while the left ovary developed, the right one shriveled up, leaving her lopsided. Gradually the germ cells reached a few millimeters in diameter, and after two or three months each one accumulated a primitive, white form of yolk inside the surrounding membrane. When she got to laying age they grew together, like a bunch of grapes hanging from the ovary.

The story of how each of these cells becomes an egg is quite exciting. One by one, each one grows larger and larger; when it is about ten weeks old, the hen's liver uses fats and proteins in her body to create a yellow yolk. This yolk then drops into the oviduct – a twisted tube, quite possibly 3 feet in length, like one of those things you slide down at the adventure playground.

If we had a cock and Loxy had recently mated with him, sperm would be waiting at the entrance to this tube, and one lucky fellow would get to join the yolk on the subsequent stages of its journey.

What happens over the next couple of hours is that the yolk (with or without sperm) spins down the slidy tube. Gradually it picks up a coating of albumen 'after the manner of a rolling snowball', as my encyclopedia describes it. Albumen is egg-white (from the Latin *albus* meaning 'white'). Once well-coated, the snowball then spins even more, twisting some particularly thick albumen into two coiled strands called *chalazae* (from the Greek for 'small lump'). The *chalazae* are to keep the yolk steady in the centre of the albumen, stretching out to either end of the shell. Like safety cords, they protect the growing chick from hitting hard shell walls before it is big and strong. Before the egg enters Loxy's womb, it wraps itself in two rough, protective membranes that stick to one another everywhere except at the wide end where they form a little air pocket.

The next stage of development is by far the longest – first the womb wall pumps water and salts into the albumen, and then when the membranes are taut like a big balloon, it slowly secretes calcium carbonate and protein to form the shell. This process takes around 17 hours. Finally, the fully formed egg goes to its waiting room at Loxy's back end where it positions itself blunt end down, biding its time until it feels ready to make its entrance into the world. It arrives from her vulva at the temperature of her body (around 41 degrees centigrade) which, as we already know, is lovely and warm.

Still with me? In truth, my five year old is beginning to lose interest, but I'm not. For me, the most interesting bits are still to come – what I want to know is, what exactly is this egg? Not from the reproductive point of view, but from a nutritional one. Over to the food science book.

Well, it turns out – surprise, surprise – that a chicken egg is an excellent thing for a human to eat. It contains about 70 per cent water, 10 per cent protein, 10 per cent fat and 10 per cent minerals. All amino acids essential to animal life are packed inside, plus vitamins A, B, D and E. It includes a plentiful supply of linoleic acid, a polyunsaturated fatty acid essential in the human diet, as well as plant pigments that are especially valuable antioxidants. The amount of protein in a single egg is equivalent to 14 per cent of a British adult's daily recommended intake; its calorific value is roughly 75 calories, three-quarters of which is contained in the yolk.

The yolk is the part we have learnt to worry about, because of its high cholesterol content. Though medics used to recommend limiting consumption in order not to increase the risk of heart disease, recent studies have shown this has little effect, probably because yolk fat is unsaturated and because other substances in the egg interfere with our absorption of the cholesterol.

Of course, keeping your own hens is one way to reduce cholesterol in your eggs, as long as you are prepared to sacrifice a lawn or two. As someone who would like not to give up all my grass, I could always get some Aracaunas whose startlingly azure eggs have the lowest cholesterol levels of all.

But they could never have no cholesterol at all. That's because the yolk is a concentrated food source for the growing chick. When you open up a hard-boiled egg, the bit that might have grown looks like a tiny white dot at the centre of the yolk's surface. This is the blastoderm, otherwise known as the germ cell. There it sits, un-coagulated, just wishing it had been allowed to become a baby chick. It is especially rich in iron.

While we are investigating the yolk, let's slice through it and examine its history. It's a bit like a tree trunk. The dark rings were created during the day, while the hen was feeding; the light ones grew during the night, while she was asleep. The general colour depends on what she has been eating over the previous 24 hours. As we know from our feed explorations, marigold-yellow yolks probably means she's been eating marigold extract; dark yellow tinged with green means she's had lots of grass and other greens; really green yolks might mean she has found a stash of acorns or the pasture weed shepherd's purse on which to forage. The greenest I have ever managed was from an excellent crop of cow parsley; it tasted a bit more earthy than usual; probably ever so good for me.

Meanwhile, the white might not have been.

Made up of 90 per cent water, with traces of minerals, fatty material, vitamins (riboflavin gives the raw white a slightly yellow-green cast) and glucose, the albumen looks innocent enough. It supplies the chick with essential food and drink; but it also plays a more malevolent role in its development. Within its unassuming form hide an array of proteins, there to protect the growing embryo. Some act like a shield against infection, inhibiting the production of viruses or digesting bacteria. Others bind tightly to vitamins and to iron, thereby preventing them from being useful to other creatures who might want to eat them.

From the eater's point of view, this makes life quite interesting. If you have ever counted your calories, then you might have considered a 'white omelette' where the fatty yolks have been removed. Little did you know that if you ate it raw, you might have lost weight at a cracking rate, as laboratory animals have done when fed raw albumen. This phenomenon seems to be to do with the most plentiful albumen protein, ovalbumin, which inhibits protein-digesting enzymes. Whatever proteins you are swallowing, it stops your body from making use of them.

Though this inhibitor seems to be deactivated when cooked, ovalbumin can still be a problem – some people's digestive systems mount a massive defence against it that takes the form of fatal shock. Which is why we are told to feed our babies the yolk, but not the white. Just in case they have an allergy.

And while we are peeling baby's egg, it's fun to examine the packaging. First, there is the air pocket between the two membranes at the blunt end. In a fresh egg it is about 2cm in diameter and set slightly to one side. If that egg is fertilised, then the membranes will loosen and the pocket expands as the chick grows, thereby maximising the amount of air available for it to breathe while it is chipping its way out of the egg, sometimes over several hours. Even in an unfertilised egg, the sack expands over time because the contents are shrinking by at least 4mg a day as water evaporates from the shell.

This is useful to us in two ways: firstly, an old egg with its loose membranes will be much, much easier to peel than younger models. Secondly, a bit like a witch trial in the seventeenth century, we can tell if an egg is rotten by dunking it in water. A fresh egg sinks; if modestly fresh its wide end will rise a little; if guilty of decay, the whole thing bobs around on the surface.

And if this test seems too unscientific, we can always get out a measuring tape and make sure the height of the air pocket is no more than the 6mm stipulated by the EU (if the space is any larger, a shop egg must be discarded). But for this we will need candling equipment.

Candling has long been the way to check what's going on inside an egg without breaking it – you place it in front of a bright light and inspect its contents, illuminated through the shell. These days commercial producers use the method to determine the quality of their eggs. Where once they would have used an expert eye and a candle, these days a scanner can do the looking and an electric bulb the lighting.

To determine the condition of the yolk and albumen, the egg is 'twirled'. The scanner is looking for a well-centred yolk, visible as a shadow only, without any discernible outline. If it can be clearly seen, its *chalazae* are not doing their job properly and it is too close to the shell, floating free, which means the egg is possibly of poor quality.

Other things sought during candling are blood spots, caused when a capillary in the ovary ruptures; 'meat spots' are similar– blood or tissue comes away from the ovary and gets incorporated into the albumen. Neither is a threat to the wellbeing of the eater, nor does it indicate that an egg is fertile. But because consumers think this might be the case, the packing companies search out the spots and discard millions of perfectly edible eggs.

There is nothing quite like the power of consumer opinion. When I was small the only eggs available in the shops were white because white-producing breeds were the best layers. But then the nutritionists came along to teach us that white bread was bad and brown bread was good. Having got that basic fact into our heads, we started applying it to other foodstuffs. Those white eggs we had been eating must be nutritionally inferior, we thought; rustic-looking brown ones would be a much healthier option.

Soon enough, the egg industry realised that the public was prepared to pay more for brown eggs. And rather than confuse us with any more complicated nutrition stuff, they decided to concentrate their efforts on producing hybrids that would lay one brown egg a day in a rustic battery farm.

Of course, the colour is entirely superficial; so much so that you can scratch it from the surface with a kitchen knife. Look – that lovely speckledy golden hue will always capitulate to white, given a good scraping. Inspect the coloured cuticle closely and you see it has a powdery bloom, there to protect the growing chick from bacterial infection. Tilt the scratched section towards the light and you will see thousands of tiny holes – those are up to 15,000 pores where gases pass in and out of the shell. As a chick grows inside, it requires increasing amounts of gas and the protective bloom will gradually break down in order to allow for this.

But if you are intending to eat the contents of the egg, you don't want that to happen. Which is why the EU stipulates that commercially produced eggs should not be washed, as washing destroys the bloom. The temptation at home is to run an egg under the tap before storing it away, but in the interests of hygiene I try not to. If it arrives covered in muck and I feel I really must wash it, then I use water that is warmer than the egg – this way, though the protective bloom is destroyed, bacteria are less likely to be drawn inwards through the pores in the shell.

Only once you are producing your own eggs do you realise the huge variety there could be on a supermarket shelf, had the industry not fallen in with consumer opinion. The genetic determination is surprisingly simple – my Warrens have a brown-egg gene; Araucanas have a blue-egg one; in a breed like the Leghorn, the absence of either blue or brown gene makes the chicken a white egg-layer.

These days there are hybrids producing all three colours of egg and variations thereof, and even a variety of sizes. You should be able to tell the general colour of the eggs by looking at the chicken – a white bird will produce a white egg; a brown a brown. If she has an odd plumage like lavender or cuckoo, then you can consult her earlobes – the little patch of skin down in the corner of her cheek, behind the beak. Sometimes the ears are hard to pinpoint, hidden under flaps of wattle. She tends to use them one at a time, so when she hears something familiar or startling – a child's cry, a dog's bark, she will stand stock still for a moment, working out from which direction it came. Then she stretches her neck and tilts her head to that side, all the better to catch the sound the next time it comes around. Watch closely and you will eventually locate the ear and its lobe. In most cases, the colour of the lobe indicates the colour of the egg.

If, like me, you stick to hens with the boring brown-egg genes, you will still discover over time the surprising variety in their shells. The most obvious variation can simply derive from the time of day it was formed. Shells are composed of 94 per cent chalk (calcium carbonate) that the hen picks up while foraging. An egg that arrives early in the morning is likely to have a thinner shell than one laid by the same bird later in the day because it was created while she was sleeping and therefore not accessing instant doses of chalk. Another cause of thin shells can be if the chicken has been doing too much breathing, either because she has a respiratory infection or because of very hot weather. As she pants she gives out extra carbon dioxide, thus reducing the amount of carbonates in her blood and leaving her without enough calcium carbonate to build a solid shell.

Eggs from old birds are the most characterful – sometimes the shell can be papery, sometimes nobbly with rough patches, sometimes misshapen. Many produce thin shells; this is because they are not managing to process enough calcium from their diet to supply the needs of their ageing bodies as well as those of the egg. Sometimes an old hen will compensate by drawing the calcium from her skeleton, leaving it brittle and weak. As they get older, they lay much less often and when they do, the egg may be very large. They can also produce double yolkers – the result of two eggs reaching the albumen producing area and the shell gland at the same time, these magnificent orbs have a ridge around the middle where they became fused.

Weird-shaped eggs or fragile ones can signal a stressed or diseased bird. Pesticides and certain drugs also have an influence (check what the neighbours have been spraying on their lawn; check the small-print on any medication the hens have been taking). If an egg arrives strangely elongated by a double band of shell at one end, it may come from a hen disturbed during the 17 long hours she was creating the shell. A helicopter, a dog barking, next door spraying the lawn ... whatever it was, her response was for her body to contract, including the muscles in the wall of her uterus. The shell inside there was very fragile and weak; the muscle contraction squeezed it and might even have cracked it. When eventually she relaxed, the uterine wall secreted calcium carbonate with extra vigour, producing this double layer of toughness.

A similar effect can be caused by internal fat pressing on the reproductive organs. As a keeper of lean hybrids, I never came across this problem, but people who keep rounder, fatter breeds, like the Orpington, say they can get striated or ridged shells.

Finally, some eggs have no shell at all.

I remember Roxy and Loxy once producing a cream-coloured blob that looked a bit like a dried pear, all squidged up. Because it had no shell, I couldn't tell who had laid it. I rang a friend who had kept hens for more than forty years: John. He told me not to worry – it was a 'lash'; a bit like the contents of the hoover; the result of the hen clearing out her egg-laying tackle. Neither chook seemed in any distress. I decided they might even feel better for the spring clean.

But it was a shock to find my first shell-less egg. Lying there in the nesting box, it looked as though it had been lightly poached and liberally seasoned with wood-chippings. When I touched it, the membrane wobbled dramatically to show how very uncooked the contents were. I carried it inside to show Jay and we buried it ceremonially in the compost bin.

What did it mean? My first thought was that Roxy and Loxy must be deficient in calcium, and I hadn't been baking enough eggshells for them. My second was to phone John again; he said it could signal sickness. He wondered, might something have frightened the hens?

Frightened them? It must have been something truly terrifying to halt 17 hours of shell production. Something actively intimidating; perhaps continually so. Something I had known about but conveniently forgotten. Until now.

Mr Fox

And nothing more was ever seen
Of that foxy-whiskered gentleman.
Beatrix Potter, *The Tale of Jemima Puddle-Duck*

There is nothing nastier for an urban hen than an urban fox. Her country cousins may have other enemies – the most common being a kite or a badger. In the city there might be the odd feral cat or a visiting Rottweiler, but nothing is so fearless nor so ubiquitous as the fox.

Surrounded by street lighting and the constant buzz of human activity, traffic and machines, he long ago lost his shyness and his nocturnal nature. There he is in broad daylight, strutting down the middle of the street. I have seen one on a summer afternoon in south London, sunbathing on the bonnet of a car. He is not just wily; he is brazen.

Before we moved in and built our shed at the bottom of the garden, there was a fox den there. The neighbours said the cubs used to come and cavort on the lawn. Ahhh. People used to leave milk out for them; one friend described in ecstatic tones how a darling fox cub came over and licked his toes (or was it kissed?) as he performed his t'ai chi exercises one morning. Ahh.

Having erected the shed, we reckoned Mr Fox had taken his family far away. The first few weeks Roxy and Loxy were out in the garden, Jay and I were concerned he might show up. I spent as much time as I could in the garden, and when indoors, I regularly checked out of the window for signs of a visit. As the months passed and no one came, I decided he had been so traumatised by his eviction that he had crossed us off his list of food-finding venues.

It was almost a year before disaster struck. The morning after Guy Fawkes Night: I remember because I had been worrying that I should have brought my chooks indoors in case the fireworks frightened them. When I went to let them out in the morning, I was relieved to see them bustle out on to the frosted lawn, just as usual. At around eleven (a time all self-respecting foxes should be snuggled up in bed) I was in my study upstairs and glanced out of my window to admire the russet triangles of my hens' upturned tails against the green and silver of the undergrowth. And there, exactly the same colour as my birds, slinking thief-like across the ground …

Call it what you will, it felt like maternal instinct that spurred me from my desk. Before you could say 'Mr Fox', I was down the staircase, out of the back door and across the grass, clapping my hands and roaring – 'Get away!' The chickens were shrieking at the top of their coloratura range. Roxy came sprinting towards me on skinny yellow legs, wings propelling her faster than Roadrunner. But Loxy was already trapped between vulpine jaws, flailing in the final paroxysms of life. As I lunged at her killer, he hopped over the fence into next door's garden.

Trust an urban fox to exploit the trespass laws – he knew I wasn't going to follow him on to Pete and Janice's property. From there he turned to face me and, adjusting the position of my pet between his teeth, decapitated her. Her fulsome body dropped to the ground hardly bleeding, no longer moving. And away Foxy trotted to bury his trophy in the bushes.

Meanwhile, having placed Roxy safely in the old rabbit hutch next to the house, I began storming up and down the lawn. But what could I do? I watched with impotent fury as Mr Fox returned from his burying ritual to enjoy his elevenses in a flurry of red down. While he munched and crunched, he gazed at me as if to say 'Who are you, you hysterical female, flapping about? I'm master round here.'

I remember the rest of that day, and days to come, how my fury gradually turned to sadness. I remember gazing out at the empty garden and feeling that things would never be the same again, our period of hen-loving innocence was gone. The children wandered around looking mournful. People say that after a burglary they feel somehow violated. That's how it felt after Loxy was taken.

As for her sister, we worried that she might perish from her own terror. For a couple of days she shuddered in her cage, hardly eating or drinking, her great laying talents gone. On the third day Ellie took her out a bowl of hot porridge laced with poultry spice and she seemed to perk up. I held her close and tried to encourage her – 'Come on Roxy, my girl. You're safe now. We won't let him get near you again.' But I felt the faltering of my tone; I knew I could not be certain of my promise. As I placed her on the ground, the whole family assembled to watch her take a few tentative pecks at the grass. That was a good sign. If she got through this, we resolved never again to leave her outdoors alone.

I think it was about a week before she laid an egg. I remember the relief when I heard her announcing its arrival – Roxy was back to her normal self. And back to the Ark I carried her, but this time as its full-time resident. I was sad that she could no longer scratch around in the open, and even sadder that she had no companion with whom to pass her days. Each time I opened the door to feed her, she tried to force her way out – in the back of her bird-brain there must have been some residual memory of free-range pleasures. But I couldn't risk indulging her.

With Loxy's death, Mr Fox had become an all too vivid aspect of our lives. I started lying awake at night, listening out for his cry. I had not been aware of it before – like a colicky baby, rending the darkness. In the morning, traipsing down the garden to check that my solitary chook was still alive, I found his calling card of a stiff black stool. Sometimes during the day I would hear squawking and know that he had come a-visiting, slinking back and forth along Roxy's wire walls. When I strode out to defend her, his shadow sauntered away – 'I'll be back,' that saunter seemed to say.

That's the thing about thieves – they know not to give up. With sheer persistence they will get their way in the end. One day some-one will forget to lock the door, or leave a window open ...

Approaching the Ark, I felt my gut tighten. I couldn't see Roxy's eager form flitting behind the mesh. As I got closer, I saw that the little door into the nesting box had fallen down, leaving a triangular black hole. Even now, I can't fathom how he did it, but my best guess is that I had not revolved the catch precisely enough to its vertical position, that somehow his snout had pushed it around and managed to dislodge the wood. I didn't want to go any closer; I knew Roxy would be gone. Eventually,

I summoned my courage to lift off the side that hid her sleeping compartment. A couple of long tail feathers clung to the ceiling, some softer plumage scattered across the floor. She must have put up quite a struggle.

And stuck to the raw edge of the nesting box was a fringe of red and white fur.

In mourning for my pets, I talked a lot about their murder. And soon discovered people in the area who have given up keeping poultry because of similar experiences. I also discovered that it was not only hens who were terrorised. Take, for example, the poor old cats round the corner; now that's a gory story. My friends Dan and Polly were clearing up the garden in the autumn when Polly picked up what she thought was a ball. It turned out to be a cat's head, severed in distinctively Foxy fashion. The eyes were still half-open, and not rotted, so it couldn't have been dead all that long. Scary. A month or so later they were woken by a growling and a yowling outside their bedroom window. The lady next door ran out screaming; she managed to chase off the fox who had abandoned his second feline victim, half-mauled. The cat, covered in bites, was nursed by owner and vet but eventually had to be put down.

And the locals' response? Well, there was one lady who phoned to accuse Dan and Polly of making up the whole story – such gentle creatures as foxes would never do such a thing. Because of this sort of opinion, other people kept their heads down, bringing their cats in at night just in case the same happened to them. Everyone brought in their shoes, otherwise the tasty leather might get mauled. One neighbour with a small baby would not leave her outside in the pram – having witnessed Loxy's murder, I thought it a wise precaution.

Not to be outdone by the fox-lovers, I began looking into ways to free us of their tyranny. I had country-living friends who depended on their chickens to earn a living. They had fox traps, and rifles to ensure a humane death. I knew I was not allowed to use a firearm in a built-up area, so I made a deal with my friend Tim to borrow a couple of traps and drive my catches over to his place, where they could be swiftly dispatched.

In the meantime, I looked into the best forms of fox-proofing.

The main thing henkeepers said, based on bitter experience, was always to be vigilant. Never assume Mr Fox is having the night off, let alone the day. You can bet your bottom dollar he is somewhere in the shadows, biding his time like the opportunist he is, just waiting for the moment when your concentration falters.

Another general recommendation was to try my darndest to think like a fox. One thing he is especially proud of is his pearly gnashers; and not only for inflicting decapitation. One morning my friend John found his garden strewn with shards of plywood – the fox had spent all night ripping the hen house to pieces.

He is also pretty nifty at climbing. Though not generally spring-loaded like a cat, he can get a long way with a messy scramble. With this in mind, I looked into anti-burglar devices. Movement-sensitive lights are meant to be a deterrent, though I doubt my savvy urban thieves would have been bothered. The Neighbourhood Watch website suggested planting prickly bushes around the perimeter of the garden or sticking spikes atop a six foot fence. I decided that rather than prune my overgrown trees and ivy, I would encourage them to clamber high and messy over the fence.

Many friends swore by electric mesh. At around £100 for 50 metres, it is cheaper than solid fencing. The reason it works is that Mr Fox always takes a sniff at a barrier before he clambers up it, and a sting to the nose should be enough to send him packing. A set of posts, 2 metres of mesh and a rechargeable battery unit is easy enough to find at the local agricultural supplier. If the posts and pegs are too weak, like my friend Elaine's, the mesh sags, sending the electric current into the ground and rendering the fence useless. Her husband went out and bought a set of treated timber which took him a few days to dig in really deep. It is now part of their regular anti-Fox strategy to make sure the top and bottom wires are taut, and the weeds are not growing up and shorting it out.

In order to prevent hedgehogs getting caught on the bottom wire and accidentally wrapping themselves around it, you might consider placing 6-inch board along the bottom of the electric fence. Once you have had Elaine's early-morning horror of finding a sweet bundle of spikes mortified on hers, you will definitely be putting one in.

Dogs are a good idea. One friend has a little terrier who polices her back garden very effectively – most foxes prefer to avoid confrontation with a member of the same order. And it is strange how dogs take it upon themselves to guard the family flock. Another friend has a Lurcher who wrecked relations with next-door when he got into their garden and finished off a couple of hybrids. Yet when my friend started keeping hens herself he became the sweetest godfather to them and later to their chicks.

Sliding further along the eccentricity scale, some people use lion or tiger droppings from the local zoo (they deter cats too).

Others find room for a pet llama. Me? First, I needed to improve on what I had.

With Roxy's theft had come the realisation that my Ark was not as secure as I had imagined. It turned out, of course, that the name 'Ark' refers to a quality not of safety but of mobility. Had I made the association not with Noah but with pigs, I would have remembered those things like Nissen huts you see stranded in the middle of muddy fields. Completely flimsy and insubstantial, they are unlikely to withstand either flood or fox.

Fortunately, my Ark was sturdy. Even so, there must be ways to make it more secure. Were I to replace all the catches with bolts, Foxy would find it much harder to enter; the addition of a padlock would also deter human thieves. However, such an upgrading would test my screwdriver skills, and henceforward my cleaning and feeding and daily egg collection would be much more time consuming. Instead, Jay offered to tighten each revolving catch – that would present a challenge for even the most persistent snout. He could also put a second catch on the little door into the nesting box so that at least next time Mr Fox came around he would have double the bother.

Then we applied our new safety-awareness to the run. We had read Roald Dahl's *Fantastic Mr Fox*; we knew our enemy was an admirable digger. To give the run that extra Fort Fox dimension, we inserted a 'scratch-mat' on the ground – a big square of wire mesh, jutting out a few centimetres beyond the walls. Though the hens' scratching would be thus inhibited, so would their killers' skulduggery.

In thinking like a fox, we tried to be equally cunning. Foxes (even urban ones) hate the smell of humans, so from now on I intended to clean out my daughters' hairbrushes in the garden, leaving fresh clumps floating about in surprising places. They could also join Jay in urinating regularly along our boundaries (Bea was particularly keen on this chore).

Or I could take a break from chicken keeping altogether. But now my fox-awareness had been raised, I was bound to find myself bumping into him.

It was 9am at the start of the Easter holidays and my neighbour Lisa was on the phone, sounding in a panic.

'I want you to solve a moral dilemma for me,' she said. 'There's a fox in the children's guinea pig cage. He must have dug his way into their run, got up the ladder, and now he's stuck.'

I saw no moral dilemma whatsoever in her news – a fox was caught; time to put my plan into action and phone Tim. But by the tone of Lisa's voice, I sensed things were not going to be quite so simple.

'He's so beautiful,' she sighed.

Uh oh. Now I knew which way this dilemma was swinging.

'In my opinion,' I spat, 'that fox would look even better hung around the neck of a beautiful woman.' She laughed; I could tell she was nervous.

The story of Jemima Puddle-Duck filtered into my mind – Beatrix Potter's broody duck, seduced by a long-tailed gentleman with sandy whiskers. Lisa and I read that stuff to our children, and despite all our better judgements, couldn't help but empathise with Jemima's fancying a handsome bastard. My friend was about to confess the real-life consequence of her weakness.

'I've rung the RSPCA and they are coming to let him out.'

'He's killed your children's guinea pigs!' I cried. Surely this was the role she wanted me to play – reminding her it was our pets that needed saving, not the fox.

'Yes, and your hens. I know if he gets Ellie's bunny you're going to blame me!'

'Did you try phoning the council pest control?'

'Yes, but they have a 'no kill' policy.'

'A what?!'

'They say foxes are a nuisance, not a pest ... government legislation; you know...'

'And what about a private company?'

'Rentokil can't come straight away. We're meant to be going to Wales in an hour. And anyway, I'm not sure I want him killed. He was only following his nature, Julia ...'

Now I object to this argument. I know about 'nature red in tooth and claw'; I respect the need to kill in many species, including humans. But when it comes to urban living, that whole nature thing goes out of the window. Urban foxes don't flourish because they are good at killing, but because we are. It's we humans whose dead chickens and pigs generously deposited in bin-liners on the street provide their sustenance.

In my opinion, the life of Lisa's handsome captive was no more 'natural' than her guinea pigs'. His urge to kill might derive from some ancient legacy, just like a dog's or a pussycat's, or even the late Loxy's when you passed her a snail, but if any of them had to kill to survive, they wouldn't stand a chance.

Reg from the RSPCA arrived in a pristine white van, bang on time. The RSPCA is generously funded by millions of animal lovers up and down the country. Which comes first, I wondered, the furry or the feathered? I decided my best option was to take Reg into my confidence.

'The thing is,' I began, 'we have a real problem around here – the foxes are terrorising our pets. It's not just the guinea pigs …'

He nodded. I wondered how Noah faced this problem, during those forty long days and nights.

'Trouble is, it's breeding time,' he said. 'I saw a cub just yesterday – six weeks old or so. It would starve without its mother.'

'But if this one's a male, there's no problem,' I responded. In my mind, these terrorists must all be male – yes, I know it's sexist, but feelings were running high.

'True,' said Reg. He opened the back doors of the van and took out a white cage, lined with a comfy cloth, and a long pole with a noose at the end. 'I'll tell you what. If it's a male, I'll take him away for you.'

'Where?' I enquired, my tone more eager than I had intended.

'Somewhere beyond the ring-road; there are plenty of rabbits up there … ' Wild rather than pet ones, I assumed.

In the back garden, the fox filled up the whole of the guinea pigs' cage. Reg and I craned our necks to try and get a good look between his back legs. A little flick of his body, and there it was – his manliness. Good. Reg reckoned he was last year's cub – maybe the son of the criminal who took my chickens. He reminded me of a handsome hoodie, annoyed that I had the audacity to look him in the eye – his golden eye, glinting out through the wire. His ears were tipped with soft black fur, his paws were clean, teeth unbroken. I could see what Lisa meant – indeed, he was so beautiful.

Reg offered me protective gloves and I became his assistant, excited and rather proud to be honoured with the role. I eased open the door of the guinea pigs' cage, while he inserted the pole, talking in low tones to keep things calm. To him, this process was no different from any of his other work rescuing traumatised animals. The fox was sniffing at the noose, picking up the whiff of cats and bunnies and all sorts of delicious previous inhabitants. Slowly, gently Reg eased it up and over the furry ears. I was thinking of my daughter Beatrice's favourite story – the little gingerbread man who is persuaded by the fox to climb on to his head, from whence he gets snaffled.

Go for it Reg, you beat this one at his own game.

At his end of the pole, Reg drew up the string. The noose was tight around Foxy Junior's neck. I opened the door wide, and Reg lopped our catch niftily into his new cage. I slammed the lid shut, perhaps with a little more vigour than was absolutely necessary.

The prisoner growled, low and throaty. Reg from the RSPCA had found him guilty, exiling him to a foreign land in punishment for his father's crimes and now his own. Lonely and confused, if he tried to come home he would face the brutality of the ring-road.

With any luck it would be a really ruthless truck.

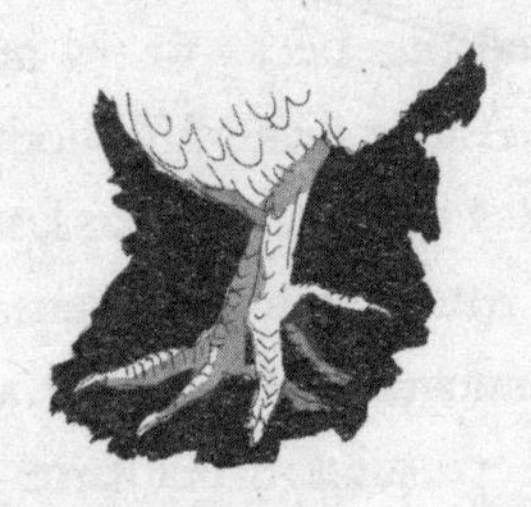

Take Two

Tomorrow is a busy day
We got things to do
We got eggs to lay
Louis Jordan

Even if Foxy Junior did perish, I knew there were plenty more like him out there. Getting hens again was sure to mean more harassment. When I couldn't bear the thought of it, I reminded myself what people say when there's been a terrorist attack. They say – get out and lead the life you want to lead; don't let the buggers get you down.

One consolation was that neighbours moving to Africa gave us their aged cat, Tigger, who might sometimes condescend to a cuddle. Unfortunately, this just made me miss Loxy's version all the more. I missed the hens' enthusiastic company in the garden; I missed their chatter round the kitchen door – *'Bop. Bop, bop, boppety bop…'* I missed the clever way they converted leftovers into eggs.

I had invested a lot of time and effort over the previous year; I couldn't let all my new-found husbandry skills go to waste. Perhaps most importantly, there was £400-worth of vacant Ark sitting at the bottom of the garden, crying out for new tenants.

After a suitable period of mourning I took advice from a wise henkeeping friend who recommended getting in touch with Mr Hodgkins, the local pullet rearer. No problem – he had a batch of POLs ready to go; £6 each.

On a sunny March morning, I drove Beatrice over to Hodgkins' place. He opened up the stable door and revealed the most marvellous sight – a hundred Roxys and Loxys jostling towards the light. Each one was sleek and beautiful, their eyes bright, chattering away to us – '*Take me, take me!*' He leant down and gently lifted the couple closest to him under his arms where they nestled happily for a moment before being lowered into their transportation sacks.

When Ellie got home from school, she christened our new pets Ruby and Scarlet and we drank to their future with elderflower cordial. The household was complete again. My pair of brand-new red hens would help me prove that henkeeping was possible, even in the same world as fox loving.

With Ruby and Scarlet came a new level of dedication in my husbandry. Like someone who has had a health scare, I now became the most fastidious of keepers.

My experience with Mr Fox meant that I didn't dare let them out without human company. Either the children would be playing in their Wendy house, or Jay in his shed at the end of the garden, the door open. Best of all, during the summer months I found time to join them in the garden nearly every day, weeding. The job was so much more fun when accompanied by the hens' chatter, or their cackle when they uncovered a grub. Between the three of us, the borders became cleaner and tidier than ever before. And at the end of a good working session, under my arms I slung them and back home to the Ark.

Having caught them, of course. With Roxy and Loxy this was not something I had had to do regularly – they found their own way home at the end of the day. But now, whenever I left the garden I would first step confidently towards each bird, hands descending. The strange thing was that, instead of making a run for it, they would cower to the ground, their feet treading back and forth nervously. It reminded me of all those 'chicken' words – chickening out, turning chicken, being chicken-hearted ... And it bothered me.

I don't think chickens are intrinsically cowardly creatures. After seeing Ruby jump on Tigger's back the one time the cat dared to aggress her, I would say they are the most courageous animals I know. I admit that the local police helicopter can send them into a huddle – their Junglefowl genes are telling them that thing wheeling in the sky is a hen-eating raptor. But I wouldn't call it chickening out. The association only seems relevant to me in this one instance of Ruby and Scarlet cowering. Apparently it is what they would have done were I a cock; I decided to take it as a compliment.

As far as handling the hens two at a time goes, my hot tip is to do it with the hens facing backwards, clenching their bodies with your elbows, their legs safely knitted between your fingers. When you carry chickens in this fashion, two tangles of claws fire out from under your arms, like some medieval torture instrument.

At markets in Mediterranean holiday destinations, the poultry keepers' style is even more extreme – they hang their birds upside down, three or four from one hand. This vertical position probably inhibits the birds' breathing and as a result they hang limp, as though their necks have already been wrung. But the general principle is worth emulating: with their bodies wedged under each arm, angle the hens so their heads are lower than their tails; even with a gentle tilt, they become pleasantly passive.

And back to Fort Fox willingly they return.

With its new scratch-mat and extra catch, plus all the family's daily deterrents, the Ark now seemed reliably fox-proof. The only problem was that the hens had to spend most of their lives inside it. Fearing they might die of boredom, I decided to treat the run as if it were a baby's playpen and introduce some amusements. The hanging ladder was already like something from a mini adventure playground, so I added a few sticks in the hope they might enjoy hopping from one to the next. Then, remembering Roxy's fascination with her own reflection, I hung some old CDs up on the mesh walls. The pigeons watched warily from the buddleia bush; they couldn't understand how those shiny things I hung all over the allotment to scare them off my cabbages were now being used to entertain their cousins.

Regular cleaning became a must for both the run and the sleeping quarters. As spring advanced and the weather got warmer, it was the best way to prevent farmyard odours percolating into the neighbours' gardens. Hens are poo-full creatures, even managing to defecate during their sleep, so it's good to have a quick and easy way of clearing up after them. Under their perch I left some pages from my weekend *Guardian*, into which I wrapped their droppings, and threw them in my canary-yellow plastic bucket, with an old dustbin lid on the top to keep off the flies. Those deposited on the scratch-mat in the run I trowelled away each week when I altered the position of the Ark.

Chicken shit is one of the most potent compost activators around; which is why garden centres charge extortionate prices for it. The chicken-keeper, of course, need pay nothing at all. Richer in phosphate, sulphate of ammonia and potash salts than any other animal manure, it is packed full of bacteria whose greatest desire in life is to break down your organic waste.

Some people put their hens' soiled wood-chippings straight on to the flowerbeds as a compost/mulch combo; apparently currants and gooseberries thrive on it. Personally, I didn't trust myself not to frazzle my plants by feeding them the fresh stuff. Instead, once a week I carried my yellow bucket down the road to the allotment, slinging its contents into the bin along with my kitchen waste. I soon noticed the improvement in my compost – even thick stems and roots began to decompose fast.

Chicken shit is not only the gardener's best friend, it is also the henkeeper's – inspected on a regular basis, it can be a good indicator of health. Fresh droppings should be firm and brown with a white surface; the brown bit is the poo and the white bit the wee. If you spot weird-coloured ones or splatty ones, if they smell acrid or mouldy, then it is time to worry.

Having familiarised myself with Ruby and Scarlet's excretions, I needed to get to know their bodies. And as every good stockman knows, the best way to do this is to sit and watch. I can highly recommend hen-watching; simply sitting with your pets in the garden is an excellent way to switch off from worries and stress. Their focused foraging, unhampered by phones and emails and all the other things that distract us humans, is an excellent lesson in life's essentials. And out of such philosophising more practical questions can arise: Does that movement look normal (no lameness or stiffness)? Is her plumage as glossy and full as usual? Are the scales on her legs clean and smooth?

At the end of a good watching session, I would try and encourage either Ruby or Scarlet to hop into my lap where, instead of my looking at her, she would look at me. And what a fascinating look it was: those perfectly circular eyes, bounded by ripples of rosy skin, the iris bright yellow rimmed with orange.

Like other aspects of her movement, a chicken's blink is extremely swift. You know it is happening but in that millisecond you can't work out quite how. Flick. The iris has been swept clean, but the folds of skin around it seem not to have moved. That's because she is using her nictitating membrane (from the Latin *nictare*, to blink). Present in birds and reptiles, this is a translucent third eyelid that enables the hen to moisten and protect her eye while retaining visibility. The rod system inside those eyes is poorly developed, so when darkness falls she can't see at all well and takes herself straight to bed. Watch her going to sleep and you get the opportunity to see the third eyelid slowly passing over the eyeball before the big ones close on top of it.

With all three eyelids open, a hen has an extremely well-developed sense of sight – strictly, her vision has only about a third of the acuity of a human's, but because of the nictitating membrane it is unhampered by those moments when we blink ourselves blind. Like all birds, she has a very wide field of view while also being able to see things right under her beak.

Having checked that my bird's eyes were bright and clear, I would continue my health check across the rest of her body – how fat was she? How full her crop? By handling her regularly,

I was in a good position to diagnose loss of weight, loss of feathers and so on. I also tried to have a good rummage around for the usual suspects – mites (around the neck) or worms (around the bum), wounds on her legs or dodgy feather loss (indicating possible bullying). And once in a blue moon, I unfolded her wing and made sure her feathers weren't getting too long.

You've seen *Chicken Run* – like the very first, primeval birds, hens cannot fly; not without the aid of a catapult. Though some of the lighter pure breeds are able to wing themselves impressively off the ground, hybrids generally don't develop the knack. Having said that, the first time Mr Fox came visiting, Roxy's full set of wing feathers probably saved her life – it was the extra propulsion of her flapping as she sprinted down the garden that got her away so fast. Of course, then I started to worry that she might improve her technique: get up enough wing power to flap over the fence, and another fence or two until she fell straight into Foxy's jaws. With Ruby and Scarlet, I didn't want to risk it, so I clipped their wings.

I knew not to worry – the quill is made of keratin, the same substance as human nails, and cutting it hurts no more than it would a baby's toes. I just had to make sure I cut the white bit, rather than the pink bit lower down where blood flows. With Jay holding on to her to stop her flapping around, I opened out one wing and used a sharp pair of kitchen scissors to cut back the ten primary feathers. I avoided the shorter feathers on the body of the wing – she needed them to keep warm. Clipping one wing only meant she was sufficiently unbalanced that even an escape attempt with a catapult would have sent her horribly askew.

Another, very particular feature of the hen's body is her comb – the wizened bit of red flesh on top of her head that looks like a discarded washing-up glove; similar bits around her eyes and below the beak are called wattles. There are human versions of both these appendages – the flesh under our chins that gradually becomes pendulous with age may be termed a wattle or else 'dewlap'. In *A Midsummer's Night Dream*, Puck mentions pouring ale down a gossip's dewlap, which some scholars reckon refers to flaps lower down her anatomy. Shakespeare is also familiar with the comb – the 'coxcomb' worn by a jester signals his wise form of foolishness.

Cocks are especially proud of their combs – they tend to be larger that the hens' and grow larger still when he is feeling aggressive or aroused. For the female, a lovely pink comb is a sign of good health; if it becomes pale then she may be ill; if floppy, she is overheated or dehydrated. A heightened reddening signals that she is going to lay. Once you know this, you need no longer be awed by people like my sanctuary lady recognising layers as if by magic. Instead, you can start impressing your friends by predicting which member of your flock is soon to produce an egg.

Ruby was a marvellous layer and to prove it she had permanently red wattles and comb. Viewed from behind, she showed other signs of laying ability: her back end had a distinctive V-shape, a wide tail area pointing sharply downwards like a super-efficient egg funnel. Her bottom was fluffier and fuller than other hens', with her tail carried jauntily, as if to draw attention to her talents.

I hope I never failed to appreciate them; neither the egg-laying nor the beauty. Certainly Ruby herself was proud of her well-feathered form. Like all birds, for maintaining her beauty she possessed a preen gland (or uropygial gland as it is more scientifically known) that produced oil to condition her feathers; the more it was stimulated, the more it produced.

A chook like Ruby knows exactly where her little sack of conditioner is stashed – turning her neck 180 degrees, she buries her face into the base of her tail. Once her beak has picked up a good supply, she strokes it back and forth across her wings, her breast and her tail. That slinky neck of hers enables her to distribute oil right down to her knickers and tight up to her collar, burrowing her face in deep to condition the roots. Discovering a louse or a flea, she pecks it swiftly from its hiding place as she goes. When eventually the preening is completed, she niftily rearranges her feathers with her beak, picking up the shafts and replacing them until they are in perfect alignment. A good shaking of her whole feathery mantle concludes her toilette.

It took some effort, but I eventually managed to persuade Ellie that the shampoo-and-blow-dry version of hen-husbandry was not necessary. It would only have upset the natural balance of oils on our hybrids' feathers. She was also wrong to think they might love us more if we gave them a hot bath. Some hens are so traumatised by water that they stop laying just because they got caught in a rain shower.

Which is why it was fine by me if once in a while my hens sheltered in the house. Just like their predecessors, Ruby and Scarlet felt drawn to our indoor life. If the back door was open, straight into the kitchen they came to have a good snoop around. One day when it was closed, Scarlet even managed to shove her way in through the cat flap and finish off Tigger's breakfast. Had I left the rest of the house available, soon enough she would have been up the stairs, finding tasty biscuit crumbs on the children's bedroom floor. I have heard of people who allow their hens to hop into bed with them – their blood temperature being a good four degrees hotter than an average human's, they could be useful on a cold winter's night. Now I am mighty fond of my chooks, but not that much.

And not so much that I forget to wash my hands after handling them. Each time the news reported children hospitalised from E.coli infection, farms closed, parents worried about ever going near farm animals again … I got the chance to nag my daughters – THAT'S WHAT HAPPENS IF YOU DON'T WASH YOUR HANDS! Though much more likely to be present in large numbers in lambs and calves than sweet ickkle chikkies, nasty bacteria like E.coli are worth worrying about, especially if you have young children. So there I stood, soap in hand, the kitchen tap running, ready for the kids to return from feeding them or fetching their eggs. As everyone who has ever been into a hospital knows, however much you adore the thing you touch, this elementary precaution is a major factor in holding back disease.

It was my born-again fastidiousness that made me take a fresh look at all sorts of health and safety matters.

Rats, for example. A fact of life in any British city, their population well exceeds our human one. They live happily alongside us, feeding on our food, using our guttering and our cavity walls for their transport system, our lofts and our cupboards for their nests. As a result of cuddly rat films like *Ratatouille*, the Great British public has recently been turning rat-lover. At the same time, with local councils reducing their pest control services and refuse collection, infestation is becoming increasingly common. From an urban henkeeper's point of view, the main problem with an infestation is when the neighbours blame it on your dirty birds.

I admit to a phobia about rats. Fine with spiders; happy with snakes and even welcoming of field mice, but when it comes to rats I turn chicken. Which was another good reason to limit my hens' access to the outside world. With its Fort Fox accoutrements, the Ark was penetrable only by the tiniest rat. Were it to have achieved this feat, it would have done so at night when the hens were tucked up, asleep on their perch. I imagined its ugly form slinking around in the dark, nibbling at the woodwork, snaffling whatever was left of dinner and finally scoffing the eggs hidden in the nesting box. Though the risk of rodents spreading disease is apparently exaggerated, I still worried about my chooks catching something from an intruder; I had nightmares about tiny rats turning into huge ones and attacking them.

In my calmer waking hours, I tried to be scrupulous. When I put leftovers out in an open bowl, I made myself sweep up any debris before dark. The hens' regular feed was in a hopper specifically designed to be inaccessible to rats; I tried not to spill any when I was transferring it to the coop. Were rat droppings to have appeared anywhere in the garden, I would have had no qualms at all about getting an electric rat-zapper.

Talking of zapping, you may have come across modern-day Rasputins who boast of their ability to lure hens into a state of trance. Generally this is done by holding the bird down on the ground and catching her attention with a forefinger held in front of her nose. By repeatedly drawing out a straight line in the air, or on the ground ahead of her, she is mesmerised into unblinking stillness. Or you can try drawing an actual line on the ground with a piece of white chalk.

I have to admit that I tried both methods on Ruby and never achieved my end. Perhaps I didn't have conviction enough to assert supreme mastery over my pet. But I did discover how to lull her asleep.

The best way was to hold her firmly in my lap, or in a comfy place like a patch of long grass and ease her head under her wing. Though it could feel as though I was about to wring her neck, I never got anywhere near. Lifting the wing wide (the one I had not clipped) I gently wrapped it around her head and held it fast, rocking her gently from side to side. Sometimes I even tried singing 'Rock-a-bye Baby'. Hey presto – she must have assumed it was time for bed and fell instantly comatose.

My second method came from watching *Ice Age* with the children – that bit where the animals play dead. The playing-dead mechanism occurs in real life as well as movies, and in all sorts of animals from pigeons to alligators – when under attack, they go into a fake rigor mortis in the hope that the predator will give up on them because he assumes they are dead already.

If you want to amuse yourself with your pets' talents in this area, you simply turn her on to her back (holding her sharp claws in to her body), stroke her gently and whisper sweet nothings in her ear. Soon enough her nictitating lid sweeps across her eyes. Once she is absolutely relaxed you can release your hold and chuckle with satisfaction at the big warm bird lying blissfully prostrate before you.

A sudden movement or a clap will awaken her.

More Eggs

I've not had one since Easter
And now it's half past three,
So, chick, chick, chick, chick, chicken
Lay a little egg for me.
Traditional rhyme

Second time around, I had got the egg timing right. While their winter predecessors had taken a couple of months to start laying, my spring POLs were delivering breakfast within days of their arrival.

'Twas ever thus. Long before Christians appropriated the symbol for Easter, eggs already meant renewal and regeneration. By the month of March the days were getting longer, the sun was rising in the sky, yet what did the land have to show for it? The seeds were hardly yet in the ground; over-wintered root crops had become wizened and inedible. Only the hens had got their act together to feed mankind.

When I was about eight I read *The Good Master* by Kate Seredy – set in the Hungarian Steppe before World War I, it told the story of a feisty city girl, Kate, going to live with her country cousin Jancsi and gradually becoming part of his family. Being a lover of animals and all things outdoors, I adored Seredy's descriptions of bareback riding and sheep herding. Her soft-focus illustrations depicted pert rabbits and ardent songbirds, children wearing embroidered waistcoats and trousers as full as skirts. Everything in Kate's world was decorative – the furniture painted with spiralling tendrils, leaves and flowers; cooking pots covered with dots and crosses. I was particularly struck by the illustration at the end of the 'Easter Eggs' chapter – two dark eggs lying together, their shells adorned with flowers, one with an athletic-looking chick tripping off its side.

> *The last days before Easter were busy and exciting ones. Father and Jancsi whitewashed the house inside and out. They painted the window-boxes and shutters in bright blue. Jancsi and Kate selected the largest, most perfect eggs, and they were laid aside for decorating …*
>
> *Evenings mother got out her dye pots and the fascinating work of making dozens and dozens of fancy Easter eggs kept the family busy. There were two ways to decorate them. The plainer ones were dyed first. When they dried, father and Jancsi scratched patterns on them with penknives. The fancy ones were lots of work. Mother had a tiny funnel with melted bees wax in it. With this she drew intricate patterns on the white eggs. After the wax hardened, she dipped them in the dye. Then she scratched off the wax and there was the beautiful design left in white on the coloured egg. In this way she could make the most beautifully shaded designs by covering up parts of the pattern again with wax before each dipping. The finished ones were placed in baskets and put on a shelf until Easter morning.*
>
> Kate Seredy, *The Good Master*

Fifteen years after I read this book, I went to do some research in the Czech capital Prague. The city was still under communist rule and had a melancholy grandeur, its rusty trams clanking, its boulevards grey and empty, as were its shops. Wandering in the narrow streets of Malá Strana in search of some memento to take home, I spotted a little boutique and made my way inside. There, piled high in baskets on the floor were delicate blown eggs just like the ones Kate Seredy had described. In dazzling colours – scarlet, indigo, saffron – were etched intricate patterns of hearts and flowers, acorns and bells. Some had pictures of couples dancing and children playing. I bought several dozen, wrapped them in layers of paper and carried them home precariously in my hand luggage.

And now, more than another 15 years later, it was time to share my treasures with my daughters. At least with the eldest, Ellie, who at six and a half must surely be old enough to be trusted with them. She and I unwrapped them together and hung them on branches of flowering japonica culled from the garden and standing in a large vase on the kitchen table. Just as we would have done were we living in Eastern Europe. And then it occurred to me that we should celebrate Ruby and Scarlet's arrival with our very own decorating ritual.

The simplest way to colour eggs is to boil them in hot-dye. Dyes can be purchased over the internet – one person I know goes to the bother of buying them from Germany; beautiful they are too. Ellie and I had not had such foresight, so we did the simplest thing and boiled Ruby and Scarlet's pale brown eggs with some onion skin. After about an hour, the shells had turned a rich burnt sienna. A rummage in the cupboard produced more dyes I never thought I had – turmeric powder made them vibrant yellow; coffee grounds, darkest brown.

Knowing that Kate Seredy's eggs had been white rather than brown, I got hold of some duck eggs from the local butcher and started to experiment with other ingredients. The bottle of cheap red wine left to go vinegary by the stove discovered its destiny as a dye that turned duck eggs stunningly violet. A handful of boiled red cabbage leaves showed up a gorgeous purply blue; spinach water made the eggs green; Ellie's favourite was the grated beetroot that came out as pink as her baby dolls' cheeks.

Natural dyes can be a bit hit-and-miss. The liquid looks much darker in the pan than on the egg, and the eventual tone of the dried shell is often a surprise. Had I been a stickler for perfection I should have boiled and strained the liquid before putting the eggs in – this way I could have controlled the intensity and the uniformity of the colour. As it was, I rather liked the mottles and blotches from leaving the onion skins in the water, and the coal-dust effect from the coffee grounds. We needed to watch over the eggs as they boiled, rolling them around in order for the whole surface to get an even chance. With some of the vegetable waters, we left the eggs bathing overnight so they could take up as much dye as possible.

In this way, Ellie and I produced a couple of dozen rainbow coloured eggs. Their surfaces were luxuriously matt in texture, and piled together in a bowl they were wonderful to behold. Our Easter holiday guests took on the challenge of eating them, dyed whites and all.

Which inspired us to turn our creativity to fancy ones. For a contemporary version of Kate's second technique – a sort of Hungarian egg batik – the simplest version is a white wax crayon and poster paint. Even Beatrice, now nearly two, was allowed a go. I experimented with wax candles instead of crayons which were OK, though for some reason beeswax worked much better than paraffin-wax. I didn't try Seredy's version using melted beeswax from a funnel – I suspect it produces the best results. If you want something more sophisticated than poster paint, you need to use cold dyes so the wax doesn't melt as you apply it. The best cold dyes I discovered were shoe dyes; some of them leave a metallic sheen when they dry. One more tip about these fancy eggs – if you want to stop people eating them after you have gone to all that bother, then make sure you leave nothing edible inside.

Blowing eggs may require a combination of a micro-surgeon's hands and a trumpeter's lungs, but it is not difficult. All you do is take a needle or a pin and carefully make a tiny hole in the narrow end and a bigger one in the wide end. You hold the egg vertically over a bowl with the wide end downwards and blow hard through the tiny hole. The contents of the egg should fall into the bowl. If they get stuck, I try sticking the needle in as far as it will go and jiggling it about – that breaks the yolk and allows it to escape.

Before I get down to decorating, the insides need washing under the tap, or else the smell of rotten egg might ruin everything. I dry them and stand them in an eggcup to do the wax drawing, and then the painting. Naturally, these eggs are extremely fragile and not suitable for the likes of little Bea (hard-boiled only for her). After I had decorated them, I threaded a string down through the middle of the egg using a big needle. With the string hanging out from the bottom, I tied off the end with a double knot and bunged up the hole with a blob of clear nail varnish. *Voilà* – another beauty for the Easter tree.

*

Despite the large consumption of eggs in our house, I never reckoned on having eggshells to spare. I needed them for my veg plot – crushed and sprinkled in circles around my plants, they kept the snails at bay; I needed them for Ruby and Scarlet – providing calcium for the next round of eggs. Apparently as I get older I may also be needing them for myself, powdered with lemon juice, to try and waylay osteoporosis. But before then, there is one other activity I use them for, because it is such fun – eggheads.

Basically, these are the empty shells of boiled eggs, the top bit removed. Wash them and dry them, return them to their eggcup and fill them three-quarters full with a screwed-up ball of paper towel, or some cotton wool. You can even try putting in some John Innes compost, which gives you the clue to what happens next: you sprinkle a little water inside the egg and then a teaspoon of alfalfa seed, or cress, or any other seed you fancy (some grow faster than others). Keep watering your mini-garden each day and soon enough you will have lovely green fronds poking from the top of the shell.

The reason they are called eggheads is that you can draw faces on the shells. You need to make sure the eggcup is small enough that most of the shell peeks over the top; or else make a little egghead pedestal with an old loo roll. Dare to use non-water-based felt-tips which are a danger to the children's clothes but ensure your egghead won't be weeping the next time you water him. Like the sugar-solution we used to use before the days of effective styling gel, his punk-rocker hairdo is also edible.

*

Before you start to worry that this is turning into a craft book, I want to move on to cookery. And before you close the pages because you have enough cookery books already, I just want to mention my grandmother's recipe book. Here it is before me, bound in linen, its spine a little torn and battered after seventy years hard labour in the kitchen. On the opening page, Granny lists the recipes by page number in her careful calligraphy, their names often referring to a friend or relation long-gone – 'Joyce's lemon pudding', 'Shouna's noodles', 'Betty's chicken dish'.

Inside, the pages are spattered and stained, the recipes squeezed up against one another as three generations have added notes and comments. Their common origin is the food culture of World War II, meaning Granny's rations plus garden produce plus fresh eggs from my grandfather's Rhode Island Reds. And once in a while, the meat from that same flock. Almost everything except chicken dishes seems to have egg in it – how to stretch your bacon? Eggs. How to make a meal out of a tin of consommé? Eggs. How to make something delectable out of yet more potatoes or spinach from the veg plot? Eggs.

It is not just defiance of wartime deprivation that pours from Granny's pages. There is also the distinctive quality of a cuisine uninhibited by calorie-counting and cholesterol misinformation. Her family expected steamed pudding for afters. They were regularly tucking in to raspberry soufflé and honey cheese pie, after they had finished off the savoury cod custard. I have a strong compulsion to set down all these recipes so that eggy pleasures once more find their place in the Great British kitchen. Pleasures that are even greater if the ingredients originate in one's own back garden. But as this is not a cookery book, I shall choose just two – one savoury and one sweet.

Mayonnaise

1 egg yolk
Pinch of salt
Half a pint of sunflower oil
1 tablespoon virgin olive oil
Lemon juice
Pepper

There is nothing so markedly home-made as home-made mayonnaise, especially when it is made with your own eggs. Because you know your hens are healthy and vaccinated, you have minimal worry about Salmonella from the raw egg. Because your egg is fresh (the fresher the better), it has only ever known temperatures between mother's body at 41 degrees and your kitchen – probably around 20. If the bowl, the whisk and the jug of oil are all nice and warm as well, then there is very little risk of curdling. Of course, Granny used a hand whisk and probably not a very fine vegetable oil. Sunflower is my light oil of choice, with a tablespoon or two of virgin olive oil at the end to give it flavour. I use a hand-held electric whisk.

I start off whisking the yolks on their own, then I add a pinch of salt, then the oil. Whisking all the time, the important thing is to add the oil very, very slowly – initially, just a drop at a time. Once the first drops seem to have bonded with the yolk, I try to get a fine, golden stream flowing from my jug. I remember my grandmother pouring from quite a height – she said the oil seeped in easier this way. Scientifically speaking, what she was creating was an emulsion of oil droplets suspended in a base of egg yolk. The water in the yolk is an essential element – the base into which the oil droplets emulsify.

Fresh egg yolk is just about the best emulsifier around because the yolk itself is a concentrated and complex emulsion of fat in water, and therefore filled with emulsifying molecules. Here we are talking lipoproteins (both low density and high density) – the sort whose levels they measure in your blood when searching for cholesterol. When raw yolk (warmish) gets beaten with oil, its lipoproteins burst out and coat the fat droplets.

All the ingredients and utensils need to be nice and warm because warmth speeds the transfer of the lipoproteins from the yolk to the oil. The salt is there from the start because it causes the yolk granules to fall apart into their component particles, making the yolk more viscous. This viscosity then helps to break the oil into ever-smaller droplets.

Eventually I am hoping for a thick mayonnaise into which I can whisk the tasty virgin olive oil, a few squeezes of lemon juice and a little pepper. The reason the result tastes so much better than shop-bought versions is that it contains no stabilisers – often carbohydrates; sometimes a kind of white sauce; check the label. A lack of stabilising agent means my emulsion may be delicious but is also vulnerable to damage from too much heat or too much cold. Up to 80 per cent of its volume is oil that can easily separate while in the fridge and will need beating back in, perhaps with a drop of warm water.

Oeufs à la Neige

2 fresh egg whites (at room temperature)
2 fresh egg yolks and one complete egg (ditto)
2 oz caster sugar
1 pint of milk
1-2 oz loaf sugar (Demerara)
Vanilla essence

This is the eggiest of Granny's eggy puddings – a sort of inverted version of the egg itself, with whites floating inside yellows. I am not sure why it is called *oeufs à la neige* – I can think of many more snowy egg concoctions than this one. Some people call it Floating Islands. Bea calls it Froth Cakes. Whatever the name, mine often turns out craggier than I intended, but it always tastes amazing.

Loaf sugar is some ancient form of sugar, pre-war even – the nearest I know is unrefined cane sugar called Jaggery from my Bangladeshi grocer. It is rough and slightly damp and sold in a block (or loaf) that you cut up. An ounce of Demerara does just fine as an alternative. By the way, the eggs don't have to be super-fresh; a very fresh egg white can be difficult to foam, but then again foam made with an old egg is unstable; eggs a few days old are the best. The egg whites need to be beaten until they stand up in stiff peaks.

Which is odd, if you think about it. Normally, beating things vigorously without heat breaks them down – butter and sugar at the start of cake-making; your children when they have taxed your patience too far. Egg whites adhere to a more Wackford Squeers philosophy – a good beating builds them up. Or, as the great food scientist Harold McGee likes to put it, 'thanks to egg whites, we are able to harvest the air.'

In his hefty classic, *Food and Cooking: an Encyclopedia of Kitchen Science, History and Culture*, McGee explains that, just like the foam on your bath, egg foam is a mass of liquid bubbles filled with air. The reason they don't collapse, he says, is due to the protein in the albumen. As the whisk pulls at the compacted protein molecules they are unwrapped; also, as air enters the liquid it creates an imbalance that tugs the proteins out of their usual folded shape. These unfolded proteins gather where air and water meet, with the water-loving bits immersed in the water and water-hating bits projecting into the air. Having thus unfolded and accumulated on the surface of the bubbles, the proteins bond with one another, creating a solid network that holds everything in place.

Of course, we need an effective tool with which to beat our foams – which is why meringues and soufflés did not enter European cuisine until the middle of the seventeenth century, when some clever chap invented whisks made of bundles of straw.

Not long after the bundles came on the scene, cooks realised they could also do with copper bowls. To understand the need for copper bowls, you have to know about collapsing foams – this is when the whole thing starts to leak and crumble. It happens because the albumen proteins have embraced one another just that bit too tightly and squeezed out the water held between. What copper does is it eliminates the strongest protein bonds that can form – those between sulphurs. Another, cheaper trick is to add acid to your egg white (half a teaspoon of lemon juice per white). It boosts the number of free-floating hydrogen ions in the albumen, making it harder for sulphur-hydrogen compounds to break up and thereby liberating the sulphurs to go off and find one another.

Even with copper bowls or lemon juice, your egg may still fail to foam. In my case, that is bound to be because the bowl is dirty. Detergent or egg yolk, oil or fat can all contaminate the process by competing with the albumen proteins and interfering with their bonding process. Even sugar can be a problem. Which is why Granny adds her sugar at the end of the whisking; any earlier and it would delay the development of the foam quite considerably.

But before we get to the sugar-adding, there is this 'stiff peaks' business. 'Stiff peaks' is one stage on from 'soft peaks', but a stage before 'slip-and-streak' where things start to collapse. This is the fullest and firmest my foam is ever likely to be – its bubbles approaching 90 per cent air content. As McGee sees it 'the protein webs in adjacent bubbles begin to catch on each other and on the bowl surface. There's just enough lubrication left for the foam to be creamy and easily mixed with other ingredients'. In other words – it's just perfect.

At this point, Granny recommends that I gently fold in the caster sugar.

Meanwhile, at the stove I put the milk and sugar and vanilla essence into a wide, shallow saucepan and bring slowly to the boil, stirring to make sure the sugar has dissolved. Then I take a dessert-spoonful of my egg white mixture, lower it carefully into the boiling milk and watch it swell. I add as many spoonfuls as I think I can safely fit without them sticking together, letting them bob up and down in the milk for a minute, then turning them over and cooking the other side for the same amount of time. What I now have is a clutch of soft meringues that I remove with a long-handled sieve and set on one side.

In a separate bowl, I whisk the yolks with the one egg and add a little of the hot milk. Then I turn down the saucepan of sugary milk so it is just simmering, and pour in the beaten yellows which I must stir gently until the custard thickens.

McGee has lots of wisdom to impart about what is going on at this stage, now the egg proteins are distributed in the milk and sugar. The sugar is a particular problem as it surrounds each egg protein molecule with several thousand sucrose molecules. With this shield between them, the proteins have to work really hard to find one another, so the liquid needs to be pretty hot. At the same time, the 'protein network' is tenuous and fragile, so just a few degrees too hot and it collapses, forming water-filled tunnels through the custard.

What McGee is telling us is that we must be patient. He compares turning up the heat to hasten the custard's thickening to speeding up the car on a wet road when you are searching for an unfamiliar driveway. You get to your destination faster, but you may not be able to avoid skidding past it, straight into curdle-dom.

My solution is to have the electric whisk standing by with which to attack any signs of the eggs starting to scramble. Eventually, having stirred and stirred and whisked and stirred, resisting all temptation of raising the temperature, eventually I have something that resembles custard. Maybe not as thick as the Bird's variety, but definitely MY birds'. I pour into a serving dish, let it cool and then bung it in the fridge so it has the chance to become as firm as possible before I place my craggy islands on top and serve.

Having kept to my promise of providing only a couple of recipes, I can't resist the opportunity to point out the convenience of eggs as pure, edible matter. As long as you have some sort of cooking facility, you can eat them utterly unadulterated; *sans* oil, *sans* sugar, *sans* salt, *sans* everything. They are one of the few things that people who never read a recipe still manage to cook. Forget pot noodles or tins of beans, a shelled egg offers the simplest form of ready-made food. For many of us, it is the first thing we choose to eat each day.

Which is why it is worth considering how to give it its best chance of success. For example, should we really take it straight from the fridge (its temperature at 4 degrees or so) and plunge it into boiling water (at 100 degrees)? Isn't that just asking for the contents to seize up and form a white so rubbery and a yellow so grainy they are well-nigh inedible? Leaving it too long or at too high a temperature can have the same results. Or else a greenish residue forms around the yolk – ferrous sulphide, created by sulphur from the albumen mixing with iron from the yolk. It is harmless, but unappetising in colour and smell.

According to Harold McGee, cooking an egg in boiling water guarantees a cracked shell. Jay tends to be the one who cooks the breakfast eggs in our house, which means he's vulnerable to a bit of nagging when I appear halfway through the process. I tell him that McGee says pricking a tiny hole in the egg before you boil it is unlikely to help; it may release a little of the internal pressure as the contents expand, but it won't stop all that knocking about caused by the bubbling water. I helpfully turn the heat down and even find a lid for the saucepan. But is this intervention appreciated?

The answer to such problems is for couples like us to agree to eschew the great British tradition of boiled eggs and have simmered ones instead. In order to coagulate, the contents don't need anything so hot as boiling water; in fact, coagulation happens at much lower temperatures, between 65 and 80 degrees. Better still, for an intact shell, a tender white and creamy yellow, we should get out the steamer (otherwise reserved for veg). Because of the coagulation thing, timings for steaming or simmering are exactly the same as for boiled eggs – soft centres need about 4 minutes; hard-cooked between 10 and 15.

Or else we can go the whole hog and leave them overnight. In the Middle East they have a long and noble tradition of being gentle to eggs. The Hamine method of slow-cooking appeals to me especially because it combines both my egg-craft and my egg-cookery skills: set some onion skins boiling in your egg pan, then lower the heat to almost nothing, slip in a couple of eggs and a whoosh of oil (to reduce evaporation) and leave to simmer for six hours or even overnight. Chocolate-hued, they are deliciously creamy.

Another version of Hamine eggs is when we go camping and want to make use of the dying embers of our bonfire. We bury raw eggs in the ash and dig out smoke-flavoured, firm ones in the morning. Buried deep enough, they don't end up like Touchstone's in *As You Like It* – 'damned, like an ill-roasted egg, all on one side.' Nor will they be broken, which is invariably the case with eggs that I try to store overnight in a tent.

At home, most of us keep our eggs in the fridge – McGee advises that it is the best way to inhibit dangerous pathogens. Though it may seem perverse of me to contradict my food science master, I have to say that the conditions offered by an average family fridge like ours are not the best for eggs. Jay and I keep them there mainly out of habit, reassuring ourselves that it won't be for long.

There are numerous good reasons not to. Firstly, in its humid atmosphere water passes into the yolk, causing it to swell and the membrane to weaken. That means when it gets to the pan, however careful I am with my 'sunny side up', I am far more likely to get an omelette. Secondly, because of their porous shells, eggs pick up whatever flavours are knocking around in the fridge. This sort of storage problem has a much greater influence on the taste of an egg than ever its mother's diet will. The garlic flavour detected in Jay's omelette, for example, is far more likely to have come from that nearby bowl of hummus than from the mashed cloves I had added to the hens' slops. Thirdly, using that set of egg-holes in the door means the delicate eggs are subjected to a regular banging and a-crashing, along with a waft of warm air from the central heating. If they are not actually damaged, they sure are unsettled.

Then there is the problem of cold eggs which, as we have seen, are no good for emulsifying or foaming or boiling. In fact, for any kind of food preparation, eggs are best removed from the fridge well in advance. I try to remember to do this, even if it means delaying supper. One day I will get the same set of wooden egg-shelves as my friend Jane, fixed to the wall near the cooker. She also Blu-tacks a pen nearby so the moment the eggs arrive from the coop, she can write the date on the shell, then be sure to eat them in the right order.

Eggs kept at a coolish room temperature (around 12 degrees) should stay edible for exactly the same amount of time as they would in the fridge – 27 days is what the Salmonella-sensitive Lion Code people recommend. The only issue in keeping eggs out of the fridge is that harmful bacteria are able to multiply much faster than they would inside it. The only way to be sure you have got rid of them is to kill them – this means simmering the eggs at 60 degrees centigrade for five minutes, or even at 70 degrees for one, or something in between if you are after a standard, soft-boiled egg.

If you are concerned about Salmonella, you should probably do this for refrigerated eggs because there are loads of bacteria knocking around in the average fridge; Salmonella is not killed by being there.

Nor in the freezer. My friend Karen sometimes freezes her excess Easter eggs. She says you need to crack the shells open in advance, otherwise their expanding contents are likely to force the issue, bursting out all over the freezer. Yolks and whites are best separated into bags or boxes, with a little sugar or salt added to the yolks to stop them becoming gunky when defrosted.

Growing up in 60s Hemel Hempstead, another friend, Bruce, vividly remembers the pre-freezer method of preservation.

'Preserved eggs!' I hear you cry. 'Pickled ones? Like trying to eat rubber balls!'

Dried ones? Far too redolent of wartime rations for Bruce's mother. Her way of preserving eggs came from an even earlier period of self-sufficiency. The same culture in which everyone learnt to pickle fruit and veg, dry their onions and bury their carrots in sand. She used a yellow, syrupy liquid called 'water-glass' or sodium silicate – a benign chemical compound used these days in things like glues and soaps and toothpaste – purchased from the pharmacist or poultry supply men. Bruce and I have managed to find an account from an old home economics book:

These early summer months are when the thrifty housewife who has her own hens, or who can draw upon the surplus supply of a nearby neighbour, puts away in water glass eggs for next autumn and winter. To ensure success, care must be exercised in this operation. In the first place, the eggs must be fresh, preferably not more than two or three days old … Earthenware crocks are good containers. The crocks must be clean and sound. Scald them and let them cool completely before use. A crock holding six gallons will accommodate eighteen dozens of eggs and about twenty-two pints of solution. Too large crocks are not desirable, since they increase the liability of breaking some of the eggs, and spoiling the entire batch.

Water glass is diluted in the proportion of one part of silicate to nine parts of distilled water, rainwater, or other water. In any case, the water should be boiled and then allowed to cool. Half fill the vessel with this solution and place the eggs in it a few at a time till the container is filled. Be sure to keep about two inches of water glass above the eggs. Cover the crock and place it in the coolest place available from which the crock will not have to be moved. When the eggs are to be used, remove them as desired, rinse in clean, cold water and use immediately.

Eggs preserved in water glass can be used for soft boiling or poaching up to November. They are satisfactory for frying until about December. From that time until the end of the usual storage period – that is until March – they can be used for omelettes, scrambled eggs, custards, cakes and general cookery.

One day, I plan to find space for more chickens and for them to produce more eggs than I can use. Come the glut, I would love the opportunity to give away boxes of them or even make a bit of extra cash by selling them to my neighbours (this is perfectly legal, as long as I don't try to do it through a shop or similarly formal outlet). I look forward to having to write the date on the eggs, store them in their own shelving, in the freezer or a charmingly traditional crock by the back door.

In the meantime, I am more likely to find that the egg holes in my fridge are empty yet again. At which point I send Ellie down the garden, reminding her of the old adage – '*The best way to keep an egg fresh is to keep it in the chicken.*'

Sick Chick

'What ought we to do?' asked Ukridge.
'Well, my aunt, sir, when 'er fowls 'ad the roop, she gave them snuff.'
'Give them snuff, she did,' he repeated, with relish, 'every morning.'
'Snuff!' said Mrs. Ukridge.
'Yes, ma'am. She give 'em snuff till their eyes bubbled.'
Mrs. Ukridge uttered a faint squeak at this vivid piece of word-painting.
'And did it cure them?' asked Ukridge.
'No, sir,' responded the expert soothingly.
PG Wodehouse, *Love among the Chickens*

A couple of months after Ruby and Scarlet arrived, we went away on holiday, leaving them in the excellent care of our neighbours. It amazed me how many people I knew were longing to exercise their husbandry skills. When I left my children for an evening with a teenager, I paid her £20 and crossed my fingers nothing happened to test her knowledge of paediatrics, let alone emergency services. Yet for Ruby and Scarlet I could afford to be choosy, picking the most responsible adult for regular, attentive visits, someone who would ask nothing in return but the opportunity to gather fresh eggs.

I knew I should not to be surprised if Ruby and Scarlet acted strangely on our return. However good their hen-sitter, they were quite likely to be hunched up and grumpy, refusing to come near. Like most pets, hens quickly get used to a routine, and it only takes a weekend away for them to go into a sulk.

Just as I had anticipated, the morning we got back we found Scarlet, the plumpest and the perkiest of the two, huddled in the corner of the run. 'Typical,' I thought, presenting her with a conciliatory bowl of scraps. By the afternoon, she was looking decidedly peaky. When I picked her up, I discovered a white discharge coming from her bottom. Damn. I stood there beside the Ark, holding my mucky bird stiffly, at arm's length, and wondering what to do next. Ruby was dodging up and down behind the mesh, eager for a run in the garden; I had better get her sister away from her in case she spread any germs.

Ideally, I should have created a sick bay close by, so Ruby could see Scarlet was still around and wouldn't treat her like an unwanted stranger on her return. But I worried about schlepping the rickety spare cage down to the far end of the garden where the fox might break in. Outside my kitchen window I could keep a watchful eye on her.

The next morning she had taken no food or water and seemed much the same as the day before, her head lolling sadly. I wondered whether I should spend £25 on a consultation with the vet and probably the same again on a course of antibiotics. And then throw away the eggs for 28 days, in order to ensure the drugs didn't get into us. I couldn't help thinking how another one of Mr Hodgkins' happy, healthy hybrids would cost me only six quid.

I got out my poultry books. In her beginner's guide, *Starting with Chickens*, the great smallholder's guru Katie Thear declares – 'not many vets understand chickens'. Good. In that case, I wouldn't bother.

But what if Scarlet got sicker? How long was I prepared to watch her suffer? How long was it right to let her do so?

I phoned Mr Hodgkins. As I stood in front of the cage, trying to describe Scarlet's symptoms, I could see that she was stretching her head upwards in crooked, straining movements.

'Could she be bearing down?' Mr Hodgkins enquired.

As someone who has gone through a couple of labours, I felt qualified to confirm that yes, this might well be what she was doing.

'In that case she's probably egg bound. Have you been away?'

'Yes – we got back yesterday.'

'Ah. Things like this do happen when you go away ... They don't like change, you know.'

'No.'

'Do you think the fox might have been round?'

'Very likely. There was nobody in the garden most of the day.'

'She seems very poorly, you say?'

'I'm afraid so.'

'Then what I think has happened is that your fox gave the hen a fright while she was laying. The egg broke inside her and now she's infected.'

'What do you think I should do, Mr Hodgkins?' There was a pensive pause.

'Have you any experience of wringing their necks?' Another pause.

'None.'

'I'm off on holiday myself in a couple of days, else I'd come round and give you a hand ...'

'Can't we get her some antibiotics?'

'Well, that'll set you back a few quid ... and it might not do any good anyway.'

I'll tell you what,' he said, 'give her another 24 hours. You never know – she might get better of her own accord.'

She didn't. Nor did she get worse. Next morning I watched from the kitchen window as her head sank towards the drinking dish, sleepy but somehow still hopeful. Ellie spent ages cooking up a special oatmeal porridge for her, with a generous topping of poultry spice for extra flavour, and set her dainty dish before her. By the time she got home from school, the whole lot was gone. I phoned Mr Hodgkins and he said hang on in there, you never know ... He was off for his annual holiday.

Next morning her head was tucked into her breast. The whole family went out to the cage after breakfast and the girls stroked her lovely golden back. We all felt wretched; I thought probably it was too late to call the vet. Only now did I realise that a hen's wellbeing is a fragile thing. That evening, Scarlet breathed her last.

What Scarlet had is not uncommon. It's called egg-binding, and once the system becomes infected it is called egg peritonitis. It is especially common in pullets – sometimes before the infection takes hold they produce a series of soft or shell-less eggs, but often the peritonitis just hits them. Once a hen starts to feel unwell from an infection like this, it takes only a very little time for her to want to give up the ghost.

Next time a chick gets sick, I will immediately diagnose the problem and turn on the kettle. Seriously. I shall have a look up her vent – is there an egg stuck there? If there is, then a cure is also within sight. Quick as a flash I will get out my tub of Vaseline, rub it around her vent then hold her over the steaming kettle (not too close – don't want to scald either of us). Or else, I shall try breaking the egg with a skewer or some such and remove the broken pieces, being careful not to leave any bits behind.

If I can't see the egg, then I have a problem. It might lie out of sight, broken (as Scarlet's probably was), and it might already be causing infection (as the white discharge indicated). Because the hen's abdomen is so close to her egg-production system, that becomes infected too. Sometimes there is no egg at all, merely a general infection of the reproductive system. Even contaminated dust entering her breathing system can be the cause, as the air sacs are very close to the ovaries.

Had I immediately taken Scarlet to see the vet, he might have prescribed a course of antibiotics and a few crossed fingers, or he might simply have offered to put her out of her misery. Poor Scarlet.

My only consolation was that at least she never lived to suffer the rest of that summer of discontent. It was on a particularly beautiful June afternoon that I noticed the next wave of trouble. My English garden was at its most glorious – clematis bells nodding from the trellis; roses rambling all over the place; yet Ruby was not in the slightest bit interested in joining me. Instead she lay in her dust tray, dipping and diving like a hippo in a swamp. I thought maybe she was trying to stay cool. She didn't look miserable exactly, and had not seemed to pine for Scarlet over the preceding weeks. I did wonder if her usually bright comb was looking paler.

Just before bedtime, I went to hook up the ladder door and found that instead of settling on her perch, she was still hanging around outside. It was then that I began to cotton on – Ruby would rather risk a night out in the cold than go upstairs to her sleeping compartment. There must be something really nasty going on in there; I went to get a torch and opened up the nesting box. There in the wood-chippings, shining scarlet in my beam of light, was a seething mass of red mite. They were crawling over one another like a pack of drunks.

I had no idea how they'd got in, these nasty little parasites, cousins of the so-much-cuddlier spider. Perhaps in the bag of wood chips, or more likely from wild birds that had scattered them amongst the geraniums where they bided their time until my unsuspecting hen came by. Like leeches, red mite can last for many months between one meal and the next.

Despite their name, they are not in fact red until they have fed, which is when they become easiest to spot. The pre-prandial version is grey and smaller because not yet tanked up, varying in size from a full-stop to a semi-colon. Mite tend to hang out in gangs, and can scuttle a few feet, but not much more. Newly hatched ones are white and frustratingly difficult to spot. If you are making regular checks for mite, as I should have been, there could well be a grey powdery frass around the edges of doors and floors, resembling ground white pepper. These are their droppings and often the first tell-tale sign that the blighters have arrived.

Their absolute favourite time of year is a humid summer because they like warm, damp conditions. Plastic houses seem less attractive to them because of their smooth surfaces and a dearth of hiding places. Wooden ones like Ruby's, with gaps between the boards (or, worse still, extra layers of roofing felt) offer plenty and have my eglu friend Fiona smirking 'I told you so …'

Though mite live in the house rather than on the hens, I should have been giving Ruby a regular search around her neck, wings and bottom. I had grown lax in my husbandry duties. Had I investigated her before I got to the nesting box, I might have found she had itched away her feathers during those long hours in the dust bath. Other signals are if the hens start to lay less, or the eggs have pale yolks and splashes of blood on their shells; some birds become sleepier and hungrier than usual. Eventually, a chronic infestation can literally suck a hen to death. Which is why it needs treating.

The question is with what? Unfortunately, most repellents merely repel; once the little buggers have invaded, you need something stronger. Grandparents who used to keep hens may recommend old-fashioned remedies that are guaranteed to work; sheep-dip amongst them. Whether they kill the hens as well as their parasites is worth considering.

Fortunately, there are effective new products coming on the market all the time, some of them '100 per cent natural' AND efficacious. At my Countrywide shop I found a canister of powder to apply to the hen herself. Made from Diatomaceous earth – the fossilised remains of plankton (or diatoms) – it works by breaking down the mites' waxy coats and dehydrating them to death.

The shop assistant recommended treating at dusk, when the vampires would be on the move, creeping out of hiding towards their prey, or simply creeping all over one another as they had been the previous evening. He also recommended that I don protective clothing – though red mite tend not to enjoy human blood nearly as much as avian, given the sort of attention I was planning, they might fancy escaping into my hair.

In spattered old decorating overalls, surgical gloves and a shower cap, I resembled nothing so much as a shabby chemical weapons inspector. Jay took charge of the canister while I held sleepy Ruby, opening one wing and then another, turning her around so not one bit of her body would be missed. With WMD all over me, my garden and my chook, I put her in a cardboard box and let her snooze there while I made my assault on her home.

If I had had a steam blaster, I would have used it to evict the intruders. Large amounts of concentrated washing-up liquid will apparently incapacitate them if you want to do that first. One old-fashioned product people risk (if they can get hold of it) is creosote. It may be illegal for domestic use but it is still very effective painted all over the coop, inside and out, and even kills the eggs (which nothing else seems to). Maybe there are some alternative wood treatments coming on the market, just as sticky and just as pungent. You have to find the hens an alternative dwelling for a few weeks, or they risk expiring from the fumes.

My weapon of choice was a spray can. The active ingredients were permethrin and pyrethrin/piperonyl butoxide, which are also in head-lice treatments; like the lice treatments, the exact balance of ingredients has to keep changing as the victims develop immunity to their powers. Pyrethrin is a derivative of the chrysanthemum flower; with piperonyl butoxide it acts on the mites' tiny nerve cell membranes, interrupting signals traveling between their brains and their muscles so they become paralyzed and die. Permethrin does the same thing and can also kill the eggs.

First I removed all the infested bedding and the perch, and swept out the inside of the Ark. Everything went in my yellow bucket to take down to the allotment where I hoped my bloodsuckers would starve to death before they discovered the blackbirds. A better option would have been to burn them – mite-infested bedding crackles most prettily – but bonfires are much frowned upon on our allotment site.

In order to reach all the mite in the Ark, Jay helped me turn it on its side and we sprayed up into the pitch of the ceiling. Like bed bugs, red mite are happy to hide up there and wait until night to drop down and feast. Well, I was not putting up with that. I sprayed all the walls where they might clamber, the perch slots where they had left an incriminating smudge of blood, every potential hiding place on floor and walls. I tried to shoot killer into every nook and cranny; I tried not to inhale. Once the whole coop was doused, I shut it up for half an hour, and meanwhile gave the perch a good going over.

Though the spray was expensive, I knew not to scrimp. On the label it said 'hazardous to bees', so I only sprayed it on the inside of the Ark and hoped no bee was foolish enough to go in there for a while. A cheaper option would have been to use poultry house disinfectant, but apparently that is even more of a threat to bees. Alluringly pink and smelling slightly of lavender, it comes in a bottle and needs to be watered down. Apart from permethrin and pyrethrin, it also contains tetramethrin (another, related insecticide), detergents and paraffin.

When the dirty deed was done I stuffed my inspector's uniform in the washing machine and took a shower. Meanwhile, Jay threw out Ruby's box and scrubbed the brushes and buckets under the garden tap to prevent the little buggers hanging out there for months to come, waiting to be reintroduced to their hosts. Despite our efforts, for several days after that first mite-massacre I found I suffered from imaginary prickly head and forearms. Out of the corner of my eye, I thought I saw red mite scuttling across the draining board or into my bed. I struggled to remind myself they were NOT immortal. They may have seemed it, but they weren't.

If you have ever treated children for head lice, you will know that persistence is of the essence. It is no good just spraying something around and then skedaddling. You can guarantee that even as you walk away more babies are happily hatching into the world, ready to take the place of the ones you have just killed. Just one survivor can become thousands very quickly indeed.

Which is why Jay and I had to do the same thing the next night, and the next, until after three days the killer contents of my can were used up. On day four we took a day off, and I went on the internet to bulk buy more spray. While I was waiting for it to arrive, I asked around for alternatives. Some friends said to try household disinfectant like Dettol and Jay's cleaning fluid (the commercial stuff, not my husband's home brand). The most convincing brew came from my henkeeping friend, Karen, who recommended Jay's fluid plus clove oil. The smell of public toilets mingled with Christmas was absolutely nauseating.

After day five's assault, I left a bit of newspaper under the perch as a mite indicator. The idea was to count the number of mite hiding in its folds each subsequent evening, but I found it simpler to open up the coop and start sweeping. If there were still red mite around, soon enough one of them would find its way on to a bare bit of forearm, crawl between the hairs and give me a nip. Horrid. And my heart sank because it meant Operation Mite was not yet over.

Which in a way it never would be. Though Ruby dared to enter her sleeping compartment even on day one of the mission, and was back to her usual, colourful self within a few days, I knew the invaders were not likely to leave completely. It was now part of my henkeeping to be fully mite-aware and make sure I always had a can of spray on hand. Though a winter freeze might shut them up and last year's infestation looked as though it had disappeared over Christmas, come the summer it could well return.

I tried to minimise mite-friendly regions. I kept fresh bedding to a minimum – just one layer of paper in the sleeping compartment and a handful of wood shavings in the nesting box. I got hold of a benign repellent powder whose active ingredient was a pungent mixture of tea tree oil and citronella and talc-ed both house and bird each week. Plus a good dose in the dust bath. And whenever the sun was shining in the right direction, I opened up the sleeping compartment – mite like things damp and dark; UV rays they cannot abide.

The reason mite are such a palaver to treat is that they nest in the house. Parasites that live on a hen's body are much easier and can, in fact, be killed en masse with anti-parasite medication. Administered through the skin, it is not licensed for sale to poultry keepers in the UK, but you can get some for pigeons from the odd website or from the vet for the cat. I had one that I squeezed on to the back of Tigger's neck to keep away fleas. The hens could have had a small dose of the same thing dabbed under their wings, and from what my online poultry fora say, this is an increasingly popular option. The price you pay (apart from the obvious expense of the medication in the first place) is you have to throw away the eggs for about a week afterwards. But for a whole three months you rest body-based-parasite free.

First and foremost, that means free from lice and fleas. The first sign of one of these common pests is likely to be a greasy clump of eggs in amongst the feathers, especially in the shelter of wings and combs. Keep searching and soon enough you find the parents running or hopping around, the size of a large sesame seed.

There is lice and flea repellent powder on the market that contains exactly the same ingredients as mite repellent, but ground less finely. A regular talc-ing should keep fleas, lice and mite at bay to some degree, but will entirely fail to combat scaly leg mite – mite that burrow their way under the scales on a chickens' legs. Some breeds of chicken like the Silkie are particularly prone to it. You know you have an infestation if the scales on her feet and legs are lifting and even falling off, with white crusts pushing up from underneath. The condition is extremely infectious and can permanently cripple the bird if left untreated. Some people recommend dipping legs in surgical spirit twice a week and then coating them with Vaseline until the parasites perish. Others recommend soaking with soapy water, drying them and then painting on the white liquid insecticide benzyl benzoate (available from vets and chemists).

Or else they save themselves the bother and give her some of the cat's flea killer, which is so powerful it kills not only all those external parasites but internal ones too. Even the ubiquitous intestinal worms.

Easy to pick up from dirty ground and from scoffing slugs and snails (damn it – I took the risk and lobbed them into the run all the same), these parasites of the digestive tract come in all shapes and sizes. Symptoms of a bad bout can mimic those for mite and peritonitis combined – the hens look scraggy and pale, their droppings may be weird colours and runny, and their bottoms mucky.

Fortunately, my hens needed nothing more virulent than a little worm powder sprinkled in their feed. The same dose a couple of times a year seemed to keep the wrigglies permanently at bay.

Until I found that I wanted wrigglies after all. My friend Paul from Zimbabwe recommends feeding hens maggots when they become crop bound. Crop binding is a very uncomfortable and quite common condition, often showing up on a summer's night, after a good day's guzzling. Sometimes the sufferer will do weird things with her head, like Scarlet with her egg binding – rolling it round to try and dislodge the offending mass. Looking carefully down the front of her body, you see something bulging out like a big breast. You have a feel – there is a lump of food stuck fast in there.

A crop bound chicken may feel miserable but still goes on feeding (after all, her tummy feels empty), which is not such a great idea. Let her go to bed as normal, but don't let her eat next morning before you have felt her crop again. If it is still in the same bulgy state, you need to try and do something about it. A common treatment is to syringe a teaspoon of vegetable oil down her throat and massage gently to loosen the lump. You can also try tipping her upside down and 'milking her' – holding her beak wide open with one hand, you knead the blockage out with the other, while being careful not to choke her. Simpler and safer than either is to find your local fishing shop and get some of Paul's maggots – down in the crop they chomp away until the lump disbands.

But they won't do much good if it's turned sour. A sour crop is when her breath starts to stink. It is probably because she has eaten some mouldy feed or too many scraps (in our case, too many sandwiches from the leftovers bucket). This time she is miserable but her crop is squashy rather than solid; she may well have diarrhoea. A teaspoon of Epsom salts dissolved in a glass of water is the traditional answer. Indeed, my *Encyclopedia of Poultry* says 'this cheap and simple physic ... clears the system of fowl more thoroughly and speedily than anything else.' If you don't happen to have any, Paul says to try yoghurt; it soothes digestive troubles in humans too. For a sick chick you make a porridge of yoghurt and oats and give it to them for a few days instead of all the other stuff. Until she is better, Paul says to avoid scraps and even greens.

But we can give her greens if she has a cold – it's the vitamin C, I suppose. If ever I noticed catarrh on Ruby's nostrils, and thought her eyes were watering, greens were a major element in my cold menu. Also, poultry spice in her seed mix; garlic squeezed into hot oats; a slice or two of fresh, raw onion in her water. Though aconite is on my list of plants poisonous to chickens, a drop of it in a teaspoon of milk used to be recommended as an effective cold cure.

Apart from the special food and drink, Ruby also got taken from her run and given a couple of days in sick bay (yes – that old bunny cage), snuggled up in a bed of straw and shredded paper. Had her symptoms worsened; had she started sneezing and wheezing as well, I would have jettisoned the folk remedies and called the vet. Chickens' respiratory systems are sensitive – you don't want them infected for long. Because she was vaccinated against bronchitis, the most likely illness would have been a bacterial infection called mycoplasma which is treated with prescribed antibiotics.

Fortunately, my birds never got anything worse than the snuffles. Which is why I never had cause to try the snuff. What Ukridge's advisor in *Love Among the Chickens* seems to be talking about with his 'roop' is a parasitic infection of the respiratory tracts. I have read that the fumes of carbolic acid, breathed in by the chicken 'to the point of suffocation' will draw the parasites out. I reckon the snuff was not quite powerful enough.

And then there are the illnesses that threaten not only *Gallus gallus domesticus* but also *Homo sapiens*. First off: chicken pox.

Just joking. Though a century ago a scrofulous condition in poultry did go by this title, these days it is known as Fowl Pox. More common in America and Africa than Europe, it is spread by mosquitoes and can be vaccinated against. As for why we call our version 'chicken pox', possibly it is because the skin of the human sufferer looks a bit like my favourite birds'. Unlike goose pimples, these ones seem to have been badly plucked. Samuel Johnson suggested an etymology I like less but feel I should mention – compared to the full-blown pox (smallpox), it was 'less dangerous' or cowardly and therefore 'chicken'. Thanks Sam.

What we are talking about here is serious stuff like Salmonella. According to the Ministry of Agriculture in France, if we all kept French Maran chickens with their glamorous dark brown eggs, we would never encounter Salmonella because the pores of their shells are so tiny that those nasty microbes can never get in.

Unfortunately, most of us don't. Instead, most of us remember Junior Health Minister Edwina Currie's gaffe in 1988 when she told the nation that 'most of the egg production in this country, sadly, is now affected with Salmonella'. It sparked outrage among farmers and egg producers, causing egg sales to plummet and Edwina to get the sack.

Ten years later the British Egg Industry Council got its act together and instituted a hygiene programme across the nation's flocks to try and eradicate the bug. In a Food Standards Agency survey published in 2004, they confirmed that out of their comprehensive sample of UK-produced eggs (tested in boxes of six), the level of contamination had dropped from one in 100 eight years previously to one in 290. This was considerably better than the figure for imported shell eggs, which can be as high as one in 30 boxes.

One major reason the Salmonella count has gone down in the UK is that commercial hatcheries mass-vaccinate their chicks against infection. Not that it has been eradicated; not by any means. It is true that vaccination considerably reduces the number of bacteria in the hens, and the damage they might consequently inflict on humans, but it doesn't get rid of them. With over 2,000 members of the Salmonella family at large (living in the intestinal tracts of most animals, including fish), they are quite likely to arrive via wild birds and their faeces. Just by pecking around her run, the average free-range hen is in danger of picking some up, and having done so she may show symptoms (diarrhoea; misery – the usual), or she may not.

The trouble is that the members of the Salmonella family that poison her don't poison us, and vice versa. So even a perfectly healthy hen has the potential to deliver a dodgy egg. Well, at least we get the chance to cook away the danger.

And then there's bird flu.

Bird flu (or avian influenza – AI) is scary. Whole books have been dedicated to its potential to wipe out the human race. Then swine flu (H1N1) came along and we all forgot about it.

In fact, the current H1N1 virus is derived from something that includes avian genes. All of the known human influenza pandemics before the advent of swine flu, in 1918, 1957 and 1968, were caused by viruses that had acquired genes of avian influenza origin: a major reason we need to keep monitoring it closely.

From a chicken's point of view, AI is one of those diseases that manifests itself in a variety of forms, many of them by no means deadly. In what are called 'low pathogenic' or LP flus, symptoms can go entirely undetected, though the annoying thing is they can mutate and become 'highly pathogenic' or HP. A duck flies into your pond carrying something unassumingly LP and next thing you know your chickens have managed to make it virulently HP.

Most likely you will know an HPAI has landed not because your birds get sick but because DEFRA gets in touch (quite possibly via someone like Lara from the council). The most likely time will be at the beginning or end of summer when migrating wildfowl are arriving in large numbers, or a shipload of infected pheasants has been imported for the beginning of the shooting season.

I have never seen a bird suffering from an HPAI, but I read that symptoms include breathlessness, a blue discoloration of the wattles and comb, oedema (head swelling) and diarrhoea. Basically, it attacks every part of a chicken's body. And having infected one bird, it will spread like wildfire through an intensive poultry farm, sometimes killing off all the inhabitants in a couple of days.

HPAIs have been around for a while. The first recorded attack was in Italy around 1878 when the disease earned itself the title 'fowl plague'. Striking the United States in 1924–25, and again in 1929, the plague was seemingly eradicated until it hit Pennsylvania during 1983–84. The twenty-first century had its first serious outbreak in the Netherlands in 2003, spreading to Belgium and Germany and leading to the slaughter of more than 28 million poultry in order to contain the virus.

'Fowl plague' is not to be confused with 'fowl pest' (otherwise known as Newcastle Disease) – a disease that used to cause havoc with domestic flocks. Like the plague, the pest attacked birds' gastric and respiratory systems; it was transmissible to humans. In recent decades it has been brought under control through mass vaccination programmes like the ones enjoyed by my hybrids.

But the plague has not. The one to worry about is H5N1, which surfaced in South East Asia in early 2004, and has led to the culling of millions of birds. Transfers to domestic poultry have occurred in France, Sweden, Denmark, Germany and Hungary ... Most likely, spread through wild bird faeces or 'respiratory secretions'.

Though vaccines for H5N1 exist and have been tried in some countries, they are not simple to administer on a large scale because they require each bird to be individually injected. The vaccine can take up to three weeks for birds to develop optimum protective immunity, and some require two doses.

For those of us who would like to vaccinate, things will not be easy. The EU Council Directive points out that vaccinated birds can become infected but not display symptoms, thus making it much easier for the disease to spread. It says that were H5N1 to hit the UK, vaccinating our flocks would not be desirable as it would increase the time taken to detect and eradicate it. However, DEFRA does say on its website that it has five million vaccines stored away for an emergency and has secured access to another five million somewhere in Spain. Probably these would be used for what is termed 'ring vaccination' – when a circle of lucky recipients is created around a central, infected farm.

The official line for poultry keepers (and their worried neighbours) is that if an HPAI hits the British Isles, we will all fall in with strict biosecurity measures like scrupulous hygiene, controlled movement zones and solid roofs over our chicken runs so a passing swan can't spit down its germs. Even for those of us possessing less than 50 birds (and therefore not on the Poultry Register), DEFRA will be instituting a strict control. And with any luck there won't be a culling frenzy.

Nor should people get in too much of a state about catching it. Virologists are interested in H5N1 – the fact that it can survive over time, in a wide variety of conditions and a wide variety of birds, means it is one of those viruses that could cause problems in future. *Homo sapiens* has not developed immunity to it. Yet according to the World Health Organisation, across the whole world there have only been 442 cases of H5N1 in humans since 2003, all in people who were practically exchanging body fluids with dead or sick poultry.

The top three countries were faraway Indonesia, Vietnam and Egypt; there were 12 cases in Turkey in 2006, but nowhere closer to Europe than this. More than half the people with H5N1 have died, though most seem to have been suffering what is termed 'co-infection' – in other words, they already had something like malaria or HIV before they contracted the flu. None of the poultry products infected with H5N1 has infected those who ate them.

Most reassuring of all is that some scary mutation of bird flu into human flu is only what DEFRA term 'a potential threat'. Even for chicken-lovers there is a significant species barrier between them and us; the virus does not easily cross over.

I called it the summer of discontent – that time when Operation Mite began. It turns out that summertime is when lots of horrible things can happen to hens – bacteria, viruses, parasites, even AI. And then comes the moult.

I was making one of my forays in the nesting box, checking for signs of vampires around the door as well as my usual egg booty, when I noticed a clump of ruby red feathers. Shades of a long-ago, cruel eviction. In trepidation, I looked into the run – there was my chook, happily hopping about. I opened up the side of the sleeping compartment and saw that below her perch there was a veritable orgy of down, as if someone had been having a pillow fight.

When I took Ruby out and inspected her, I saw that most of her back and neck feathers had fallen out. I couldn't resist a gentle tug at the remaining few – they slid into my palm, leaving her with a perfect collar of scraggy skin. Over the following week the balding process extended down her breast and eventually even her fluffy tail was gone. Had not her neck feathers already started growing back in a soft pinkish down, she would have looked like a walking oven-ready.

The reason the moult happens to hens is the same as for wild birds – they need to renew their plumage, ready for winter. It normally happens to hybrids when they are about 14 months old. Depending on what time of year they were born, some get going as early as May, others like my friend Fiona's wait until November, when it really is rather chilly for undressing out of doors.

Feathers are made of much the same stuff as eggs – minerals and protein; a chicken should not be asked to produce both at once. This is the reason commercial farmers cull when flocks are only one year old – they can't afford to hang around, waiting for their birds to don new outfits.

Having said that, such was Ruby's dedication to laying that during her moult she hardly faltered. Given some TLC and a tonic or two in her drinking fountain, soon enough her pink down had turned russet. And well before the cold weather hit, her full plumage was back, even more warm and glossy than before.

007

Higgledy Piggledy my black hen,
She lays eggs for gentlemen.
Sometimes eight and sometimes ten.
Higgledy Piggledy, my black hen.
Traditional rhyme

When Scarlet died with a broken egg up her bottom, my daughters were absolutely fascinated. The way the corpse went from warm to cold; the way her eyes sealed over and her claws curled. There she was – a really stiff bird; absolutely dead.

Death – the last taboo – that great mystery we twenty-first century folk seldom speak of and seldom experience first-hand. Not until it's our own. What Scarlet offered the children was an introduction to this mysterious phenomenon from the safety of their own backyard.

But what to do with her? Strictly, the council don't allow burial of pets in the garden; they say put the body in a plastic bag and chuck it in the bin. Ellie wasn't having that. She wanted a proper burial; and I rather agreed with her, whatever the council's rules.

We didn't have a whole load of room in our garden, not with the deliberately untended trees and shrubs, and snail-bearing pots all around. In the end, we decided that the most appropriate site would be at the feet of her sister Ruby. In case there was any possibility of contaminating the ground, Jay dug a deep hole – more than a metre down. It looked like a real grave, especially towards the end when the gravedigger had to complete excavations from inside, with only his shoulders visible and showers of soil shooting up over the side.

In slid the body, off the blade of the shovel, into its vast burial chamber. And then came the funeral ceremony, complete with Ellie's poem inscribed on a piece of card cut from the Weetabix packet. She had spent ages embellishing her text with daisies and tendrils – a six and a half year old's version of an illuminated manuscript.

'*Our Scarlet, who art in Heaven, Hallowed be thy name ...*' she recited, in high solemnity.

Then Bea and she ripped magenta petals from the rugosa bush and anointed Scarlet's corpse with them. When eventually their consecration was complete, Jay and I followed with handfuls of earth.

'Rest in Peace,' I said, my throat catching at the thought that I should have taken her to the vet.

It was Ellie's idea to sing a hymn, though she wasn't sure what. My suggestion seemed to fit the bill:

'Our Poor Bird, stay thy flight
Far above the sorrows of this sad night.'

Admittedly, the real-life Scarlet could never have flown even as high as the neighbours' fence, and our ceremony was taking place not at night but just before lunch. Nevertheless, a bird lament in a minor key felt absolutely appropriate. Up and up the scale it flies until on '*above*' the melody has risen as far as it can bear, only to meander its way sadly back to the tonic. We sang our dirge in a two-part canon, Jay and Ellie on one part and me on the second, closely echoed by an enthusiastic squeaking from Bea.

Not yet two years into our henkeeping and already we had lost three chooks. This one felt the worst – perhaps because it marked a humiliating hattrick, or perhaps because the body was uncomfortably still in sight, rather than making its way down a fox's alimentary canal.

The ceremony concluded with Jay's shovelling the soil back into the hole and all of us stamping it down in a merry death dance.

It is not good to get sentimental about hens; Mr Fox had taught me that. Scarlet had taught me that even the birds themselves have a tenuous hold on life – one egg awry in the daily production line, and a healthy pullet can become just another lump of chicken flesh. From now on, I resolved, I was going to be a much harder-hearted keeper.

Already my daughters' enthusiasm for henkeeping had begun to wane. Beatrice had grown out of the tail-grabbing and the '*chick chick*' chasing. More often than not these days I was scattering corn alone. And Ellie – well, it was as much as she could to do to fulfil her duties to the rabbit. As for egg-collecting, that had lost much of its charm when the hens moved to the bottom of the garden – just that bit too far to traipse in pyjamas.

So – the chickens might be good fun for a funeral, but they were not really the children's. They were Mummy's mini farmyard. They might be picturesque and charming; they might be affectionate and even surprisingly savvy, but they weren't pets. The reason I kept them was that they were good at transforming leftovers into eggs.

My friend Aly had given me some poultry magazines dating from her childhood enthusiasm for keeping chickens. Printed in the 1960s, they were full of black and white photos of smiling, besuited men wielding pipettes and microscopes, measuring tapes and weights. Their slogan, printed bold and red across the page, read 'Science is the answer for Poultry Profits'. Attributes under development were 'feed conversion and liveability'. Already they had managed to eliminate broodiness, to have 'minimised cannibalism' and to have created hens 'with a temperament suited to modern production methods'.

Aly's mags mark an interesting stage in industrialised chicken farming. It was during this period that fabulous Higgledy Piggledy in the nursery rhyme, able to produce eight or ten eggs at a time, had started to become a reality. In 1900, average commercial production in the UK had been 83 eggs per hen per year; in these magazines it had risen to around 260. As if they were designing a fast car or creating a computer system, the breeders were hugely excited about the possibility of ever-greater efficiency and productivity. And over the next half century they did pretty well, raising the average figure another 50, to over 300 eggs per bird per year.

Another interesting feature was that the hybrids in the magazines had names like 'the 606', 'the 505' and 'the 404'. They weren't called 'Bluebelle' or 'Speckledy', as they would be these days. Once you have faced the fact that she exists primarily to serve a factory production line, a name like that starts to seem a bit absurd. Next time, I was going to call mine 007.

When I told my experienced henkeeping friends I was planning to get another pullet, they said, 'Why bother? Why not leave Ruby on her own?'

I said 'But the Ark is big enough for two or even three. My family eats a lot of eggs …'

They said, 'If you introduce a new hen, Ruby is bound to bully her.'

And somehow, at the mention of bullying, I found myself sucked into my old, sentimental ways again.

'Hens are sociable creatures,' I heard myself cry. 'I don't want Ruby to be lonely!'

The friends explained that an intrinsic characteristic of hendom is the strictness of her social structure. Even in a flock of only two, the pecking order can be immutable; a newcomer throws everything into disarray.

'Don't let people tell you hens are all sweet and sociable,' said my friend John, whose decades of henkeeping had taught him all there was to know, 'they're not; they're vicious!'

And whatever husbands may claim, no human version of hen-pecking will ever be as cruel as its avian original. It isn't necessarily a foreigner that the pecker picks on; sometimes she attacks her companions, yanking out their feathers, often at the tail. She may be doing it out of boredom, especially if the flock is confined to a small run or cage. She may be doing it because something has changed in her life and she is feeling stressed. She may be doing it because she is a natural-born bully.

Though not nearly so common as anti-social cocks, anti-social hens can attack not only their peers but also their owners. When you enter their territory, they peck at your legs, or, worse still, when your back is turned they fly at you, beaks and talons bared. There are various tactics for bringing them to heel (including holding them upside down; dunking their heads in cold water; giving them a good, firm kick into the air …) As someone who expects even the rabbit to show respect, I wouldn't be averse to the most brutal of these if my hybrids were attacking me. Fortunately, those clever 60s scientists got there first.

John said he once came across a flock of Black Leghorns who started heavy pecking during their moulting season; he thinks it was the natural shedding of feathers that first gave them the idea. Leghorns can be pretty aggressive. Previous to this they had nibbled at one another's neck feathers, but never a drama like this, yanking out tails and wings.

John thought perhaps it was a sign they weren't getting enough protein in their food so he recommended their keeper make up a deliciously nutritious meal with a bumper pack of cheap margarine and an old loaf of multi-seed bread. They wolfed it down and went straight back to their cannibal ways.

The keeper went back to the larder and got out a pot of English mustard. An old remedy – the sharp-tasting mustard goes on whichever parts of the victims' body are being mauled. He daubed the yellow goo all over their tail and wing feathers, which must have made them feel even more warlike. A week and two large pots later, they seemed to have frightened one another into submission, which was a great relief, as even in those days mustard wasn't cheap.

It was only when things had settled down again that John and his friend realised it might have been the hot weather. Apparently, a sudden heat wave can set hens a-pecking. These days, organised people buy in some 'anti-cannibalism spray' for the summer, just in case. It contains paraffin, ethanol and other nasties, so you must remember not to light a match near anyone that's wearing it.

With some flocks, and especially confined ones, a hen-pecking problem can reach grotesque proportions. They might be attacking a bird that is sick, or a newcomer to their pecking order. In my friend Jane's recent case, it was both.

She was very surprised when she found that Poppy was broody – she has kept half a dozen hybrids for a few years now and, like me, trusted their breeders. She tried all the usual tactics – forcing Poppy out into the run; locking her out of the coop until it got dark. But there was only one place her chook wanted to be – inside the nesting box, nesting.

Jane has two very young children, plus a dog, plus a husband who is out from 7am until 7pm every day. She really didn't have time to get hold of fertilised eggs, let alone look after Poppy and her chicks. Her modest garden was already bursting under the pressure of six busy hybrids, the kids and the dog; there was no room for more.

Just as Jane was getting to the end of her tether with Poppy, a friend phoned who happened to have a cock in her flock.

'You don't fancy borrowing my broody?' asked Jane. And lo and behold, the friend accepted.

Poppy was gone for nearly three months. When she came home she was skinny as anything, having forgotten to eat while she was sitting, but mighty chuffed to have performed her surrogacy so well. Jane was happy to have helped her friend create a brand new batch of pullets. What she had failed to anticipate was the hen-pecking. Three months is a long time in a chicken's memory; so long that the other members of the flock had forgotten all about Poppy. The moment she landed in the run they ran at her, screeching. Immediately, she turned tail, only to find the gang following close behind. Round and round they chased, pecking at her emaciated body. She was too weak to stand up for herself, and maybe a bit dazed after the bliss of all that mothering.

Having stepped in and rescued the poor bird, Jane brought her indoors for the night. Next day she tried putting Poppy in with them and they came at her, so instead she left her to peck on the lawn. She thought maybe the other chickens would see her through the mesh and start to get used to her.

That went on for a few days until one evening, busy with bedtime, Jane forgot to bring Poppy inside. The moment she heard the screeching she knew exactly what was going on and ran outside with a stick. The fox ran off, but Poppy's wing was drooping and her tail feathers were gone. Next morning, she seemed surprisingly perky; Jane reckoned she had survived her ordeal and thought it best to continue her attempts at reintegration.

That was her big mistake. In retrospect, she says, she can't think why she didn't stay and watch what the hens were up to. In her defence, I remind her she had the school run to do. She wasn't gone for long – an hour at most. When she came home she met the most horrible sight – Poppy with her guts hanging out. The flock had pecked her to death.

I did wonder whether it was worth getting a new hen, with or without a name. I tried to stop myself thinking soppy thoughts like – Ruby would never bully another chicken; I can't make her suffer a solitary life.

Eventually, my friend Karen came up with a solution that adhered to the concept of hens as anonymous egg-producers while still indulging my more sentimental nature: she suggested I adopt an ex-bat.

There is no creature quite so anonymous as a battery hen. In order to produce an average annual yield of 338 eggs, she is kept in unremittingly midsummer conditions – the heating turned up to 22 degrees centigrade, artificial 'sunrise and sunset' provided by electric light. Her gloomy shed is stuffed with 20,000 birds or more, sometimes up to 100,000, in cages six stacks high.

The trouble is, she is not just an egg-machine. Cooped up four or five to a cage with nothing to do, she takes out her frustrations on her fellows; she would peck them to death had not her beak been removed. If she doesn't die from disease (more than 10 per cent do) this angry, moth-eaten creature survives approximately 72 weeks before, crippled from confinement, she is regarded as 'spent'. At which point, people like Karen step in. Rather than becoming a stock cube, the hen is taken to a loving home and spends the rest of her days scratching around in someone's back garden.

There is a registered charity, the BHWT (Battery Hens' Welfare Trust), who organise a national network distributing ex-bats to new owners. There are also individuals who chat up battery farmers and procure them for themselves. That farm is never hard to find – just follow the pall of ammonia.

Karen lived near one such farm and had a deal with the farmer's wife to pass on the healthiest birds due for culling. She reckoned an ex-bat would be the perfect new companion for Ruby. Her theory was that someone who has spent her whole life in a chick-eat-chick world is bound to know how to stand up for herself. If she proved a less superlative layer than she had been in her cage, then I wouldn't mind – I felt compensated by the satisfaction of rescuing her.

I was aware that I would not necessarily have such an opportunity in the future. After more than half a century producing the majority of British eggs, the battery industry is in for a radical overhaul. In their wisdom, EU legislators have drawn up new rules and, as usual, the UK authorities are keen to institute them. As of January 1 2012, battery farmers will be investing in 'enriched' cage systems that provide larger, more comfortable spaces, nesting and roosting facilities and even dust baths. From the chickens' point of view, their prison will start to look more like a modest B&B, and they might not want rescuing at all.

When 007 arrived she resembled nothing more than an escapee from a concentration camp. She had no plumage on her breast and hardly any on her back; her beak was mangled like a bad hare lip. One of the most common things that happens when battery hens are removed from their cages is that their legs get knocked, and sometimes the wings. Arnica cream worked wonders. Occasionally (though fortunately not with 007) she might have a broken bone and then you have to ask the vet to set it – apparently wings and legs can be mended quite successfully.

The main issue on our ex-bat's arrival was where we should put her. Placing such a creature straight in with Ruby seemed like asking for trouble. Anyhow, the general rule is that a new bird should be kept in isolation for two or even three weeks in order to make sure they don't have anything nasty to pass on. I possessed the perfect quarantine – once a home, then everyone's sick bay: that cage I had salvaged from the dump. Where better to make an ex-bat feel at home? While she was there, I got the chance to discover red mite and douse her with killer, and she got the chance to rebuild her strength.

Karen had brought a small bag of layers' mash with her. The BHWT strongly advises feeding this to ex-bats at the beginning of the adoption process – it is what they ate throughout their caged life. They also advise that we resist the urge to spoil our refugees with treats – no more than a dessert-spoonful of Cheerios each day, otherwise we run the risk of upsetting their tummies. From my feed supplier, I got hold of a sack of Ex-Bats' Crumbs that are even higher in protein, vitamins and minerals than the usual ones.

It was early September when 007 arrived; the nights were warm so I didn't need to worry about her adapting to the vicissitudes of outdoor temperature. After living in a smelly sauna all her life, her comb was large and floppy (it had acted as a heat dissipater). Gradually over the first week she was with us, it reddened and shrank to its normal size.

She had been de-beaked – the top section burnt to a stump. I reminded myself that, like claws and feathers, the beak has no sensation at its far end. It was very satisfying to watch it grow into the sharp and symmetrical utensil it was always meant to be.

After a week in quarantine, her beak and feathers were already starting to grow and she had perked up enough to venture out of her cage. I carried her down to the end of the garden and let her practise her pecking skills in a corner beside the Ark. Having spent her whole life in a cage, she might have found the whole garden or even the run an overwhelming experience; this way, under my close supervision, I hoped that she would feel safe while Ruby got a reminder that she was not the only chicken in the coop. Not that Ruby seemed in the least bit interested. After too many non-introductions through the sides of the run, I decided the time had come for a real meeting.

There are various things you can do to minimise the trauma of introducing a new hen. One is to establish the low-status newcomer in one part of the garden and then bring the leader of the flock into her territory for a few days. Having made friends on the new hen's patch, the leader should forget to bully her when they go home together. Like Jane, my problem was lack of space – I had nowhere that my birds could fit together, except in high-status Ruby's run.

The recommendation I *could* put into practice was to sneak 007 into the nesting box after dark, when the lonesome Ruby was already fast asleep on her perch. Early next morning I let down the ladder from the sleeping compartment and opened up the sides of the run so they could escape if need be.

First came she whose territory it was, her hackles raised, strutting about and squawking,

'Get out of my house!'

007 proved just as feisty as Karen had anticipated. Stumbling downstairs with her bare breast high, she expanded what neck feathers she had into an impressive ruff, flapped her wings hard and screamed,

'It's mine too!'

Ruby shrieked off out of the run, pursued by her fearsome night guest. When Ruby turned around and tried to peck her, 007 made even nastier jabbing movements of her own. When she jumped on her, talons bared, the ex-bat retaliated in similar style.

Perhaps Ruby's being on her own rather than part of a gang was the reason things turned out so well. Definitely her foe's superior fighting skills were an important factor. After half a day of

Hitchcockian activity, they decided pecking at the lawn might be more fun.

The friendship that evolved was glorious to behold. Ruby must have realised from 007's scraggy looks that she needed lessons in basic life skills. Having been hothoused so literally, she had never learnt to perch, or to lay eggs in a nesting box or to eschew the cold, the wet and the wind. It was Ruby who demonstrated how to move away from the sides of the run during a rainstorm, how to clasp her feet around a wooden rod and balance there all night long, and how to drop her eggs in a pile of wood shavings rather than wherever she happened to be walking.

By the way, it doesn't tally that an abused bird develops an abusive personality – her feathers restored and her bones toughened, 007 proved herself an excellent weeding companion. Those cheery chaps in Aly's magazines did a good job – the genes they so carefully selected to withstand battery imprisonment are just the ones that make her a good gardening assistant.

I wish I could conclude 007's chapter with that satisfyingly happy ending. But I would not be telling the whole story. Though healthy and affable and generally a successful addition to our household, she did have one very nasty habit.

Egg-eating.

It took a while to realise what was going on, as she was very good at cleaning up after herself. I thought the lack of eggs in the nesting box was because she was giving herself a break after her period of slave labour, and that Ruby's productivity was going down as the days shortened. But then one day – a half-eaten mush. I was shocked. Such an unnatural activity, I felt; shades of infanticide – gobbling up the very thing to which you have just given birth.

I blamed myself; I thought probably the shells I was crumbling in the feed had received too little baking and the hens had developed a taste for raw egg. Then I blamed the battery farm – their cages have sloping sides and collecting bays where the eggs roll far out of reach; an ex-bat might turn infanticide because she was so excited to have access to them at last.

Back to my luxury cage 007 returned, plus a packet of mineral supplements just in case eating eggs was a sign that she needed even more vitamins and minerals than those provided in her feed.

My plan was to try to stymie her fun by rushing out and removing her eggs as soon as they were laid. But I had no idea what her laying pattern might be; probably it was far from regular, with days between one egg and the next. Anyway, even on the occasions when she produced something, I wasn't necessarily going to be around to whisk it away.

Someone recommended filling empty shells with chilli or mustard, taping them up again to look like normal ones and leaving them in the nest. It sounds revolting, but often the infanticide hen is so far gone that she eats them all the same. Someone else recommended my putting in rock-hard pot eggs instead. The ones I had bought for Roxy and Loxy were sitting, forever mid-boil, on Bea's toy cooker. I reclaimed them and stuck them in the cage – with any luck, a bruise to that newly-restored beak was exactly the lesson she needed.

Meanwhile back at the Ark, proof of Ruby's innocence arrived each day in the form of one untouched egg.

And then one morning, about a week into her confinement, I heard 007 calling,

'Bra – not, not, not, not, not, not; Bra – not, not, not, not, not, not!'

Out to the cage I rushed and there, lying in the chippings – three whole eggs. I couldn't tell whether she had tried to attack the false ones, but the presence of that extra real one, entirely uneaten, was proof enough that my ex-bat had relinquished her habit and should be allowed back into the Ark.

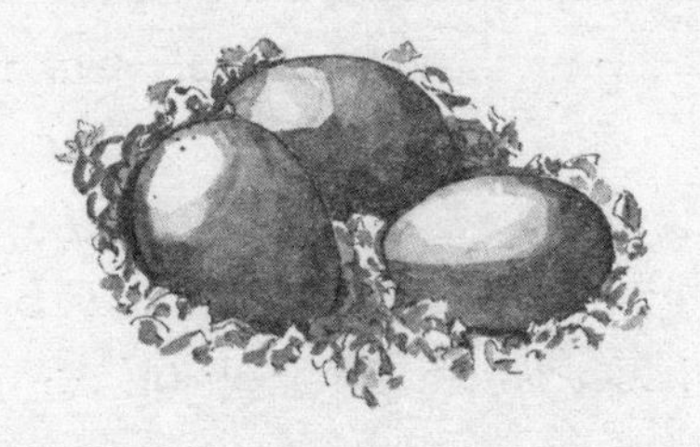

Headless Chickens

It was a dish for old Caesar,
Also King Henry the Third,
But Columbus was smart, said 'You can't fool me,
A chicken ain't nothin' but a bird.'
Cab Calloway

From the summer Ruby and 007 got together, life settled down to a regular routine of cleaning and feeding and laying. I tried to keep my husbandry to no more than five minutes a day. I tried not to fall into old sentimental habits like letting them into the kitchen or calling them by name. I tried to be vigilant about egg-eating and mite and never giving them the freedom of the garden unless I was present. My second season of henkeeping came and went without another visit from Mr Fox, as did my third and then my fourth.

At last, we seemed to be living a reasonably unchaotic version of the Good Life.

As the summer holidays came round again, we planned a few weeks away in Cornwall. My hen-sitters were keen to oblige, but we also had the cat and the rabbit to think of. For £9 a visit, Kylie from *Pets Plus* promised to cuddle the cat, and give the rabbit and the hens a good run in the garden. With that much activity going on, I thought, the fox might assume the whole family was still around and not bother harassing our animals.

Ellie (now nine) and Bea (five) were just landing their buckets on the beach when Kylie called. She was using that tender tone of voice commonly favoured by funeral directors … 007 had died peacefully in her sleep.

I was thinking, 'She was only four and a half!'

What I said was, 'Just chuck her in the bin!'

Kylie's silence suggested my response had not been appropriate.

'It's what the council recommends,' I added.

So now I knew. It turns out that most hybrids, especially those that have slaved in a battery farm, are unlikely to last more than five years. Roxy and Loxy would certainly never have survived the 12 years their sanctuary saleswoman predicted. The word 'spent' says it all – it's their hectic laying programme that kills them.

In fact, most laying hens in the UK (free range, battery, barn … whatever) are culled much earlier than that – at one year of age, before they become too susceptible to disease, let alone reduce their laying capacity. Even the happiest free-range birds are likely not to have a good death – packed into trucks, they travel hundreds of miles to one of a handful of abattoirs where they are gassed en masse. Their carcasses are not saleable as conventional meat, though some might recycle as stock cubes and cat food.

In contrast, outside the industry, traditional breeds live much longer. According to the *Guinness Book of Records*, it is an Old English Game Fowl that has so far enjoyed the longest chicken life. Matilda belonged to magicians Donna and Keith Barton from Birmingham, Alabama, and took part in their show for over a decade. Her job was to appear from under the lid of a pan into which Keith had cracked an egg. Ironically, she herself never laid one and it is probably this conservation of effort that allowed her such an extraordinarily lifespan: Matilda died in 2006 at the age of 16.

When we got home from the seaside, Ruby was scuttling up and down inside her run. She seemed as unperturbed by 007's departure as I was. But as the days shortened, her egg production went down. And down. By September, my reliable one-a-day bird was laying only three eggs a week, large and sometimes fragile. She looked happy enough, was eating and pooing and pecking perfectly normally, but I decided the reduction in eggs was a sign she was pining. Many keepers more seasoned than myself say that hens pine for their fellows; after three years of companionship it would be only natural. She had just taken a while to realise 007 wasn't coming back.

The prospect of getting a replacement was not good. Frankly, I couldn't be bothered going through the whole quarantine process again just for 007 Number Two to fall off her perch. Trying to integrate an autumn pullet (if Mr Hodgkins the rearer had one) would mean loads of hassle and possibly very few eggs for several months; better to wait until the spring. In which case Ruby would have to get through the cold, cold winter alone; with no companion to help keep her warm.

I did the maths. She was four years old; if she lived another year (which she might well, having not had 007's crummy start in life), at this rate she was going to produce a maximum of 150 eggs. Meanwhile, I would be spending about £40 on her feed and treatments – more than those eggs would cost at Sainsbury's. Add on the time spent cleaning, feeding and watering her and I was losing money.

The hardened keeper in me knew the solution – my lonesome bird should be culled.

My friend Elaine once told me that when her eleven year old learnt it might be necessary to kill one of their chickens, he burst into tears.

'Does he eat chicken meat?' I enquired.

'Yes,' she admitted.

'Then he should learn that it entails killing – if we're not prepared to kill, then we shouldn't be eating meat ...'

She sighed and looked away. She was thinking what a ruthless bitch I was, or cow, or ... is there a word for this that is not a female animal? She felt I did not appreciate that special emotional attachment between a person and a pet.

And now the pet thing had returned to haunt me. Long ago, after my mortality hattrick, I had decided that hens were not pets but egg-producers. With my last surviving hybrid beginning to lose her ability in that area, I was in a dilemma.

What if I accepted that she was my pet? People spend a lot of money on their pets; without an animal National Health Service, it sometimes proves more than they spend on themselves. Even when Ruby stopped laying completely, I might feel she was worth it because of all she had given me in the past. But for how long? As she became increasingly decrepit and in need of the vet, how much would I be prepared to spend? What if she ran out on the road and got hit by a car?

Here's an interesting fact that might have something to do with all those jokes about chickens crossing the road: *Gallus gallus domesticus* has no right of way on the public highway. Ducks and geese do, but she doesn't. If someone ran over my hen, he or she would not be obliged to pay any compensation at all. Even if the surgery cost thousands.

Enough. It was time to alter my perspective. I needed to concentrate not on my needs but on Ruby's, here and now. Was it kind to keep her on her own like this? Might it not be better to put her out of her misery?

My cowardly self was inclined to leave her out for the fox. He would dispatch her in a day or two, cleanly and efficiently, without my having to witness anything but the odd floating feather. My responsible hen-husbandry self said – No. Death by fox is seriously traumatic. I certainly wouldn't want to spend my final moments of life being chased by that ruthless bully, snapped between that set of fangs. Nor would I like to endure a trip to someone else's place for some sort of impersonal, professional death. The local abattoir was probably the Dignitas of the hen world, but if the death were mine I would prefer it safe at home, in the arms of my carer.

Which meant me. It was my garden where she had enjoyed a luxurious free-range life; a better one than a free-range commercial chicken, let alone her battery cousins. As the person who had granted it I felt I was obliged also to give her a better death.

I thought back over the years – had I ever killed an animal? Plenty of slugs and snails under my boot; the odd fish or two. I had held plenty of sick and dying birds and mammals in my arms, but had never actively brought about their death. Though I remembered firing a rifle at tin cans as a child, I had never aimed one at a pigeon. When a baby blackbird had recently been half-mauled by the cat, it was Jay who took responsibility for chopping off its head with a spade.

Now, in my 40s – already at the age when expert killers start losing their sangfroid – I was planning to dispatch Ruby. With my own bare hands.

'Argh!' Cried my daughters. Though fascinated by the slow demise of Scarlet, they were horrified at the thought of murdering her sister.

YouTube was chock-a-block with home videos demonstrating that the best way of killing a chicken was to chop off its head. I had even seen it in real life – my friend Fred who had been breeding happy meat-birds for half a century had an axe block just outside his back door. One of his flock had earned herself the name Anne Boleyn because she so often offered herself there.

I knew from Fred's set-up that the axe method could get very messy, with the dead bird flapping horribly. The prop he used to prevent it was a 'killing cone' which is basically a traffic cone with the pointed end removed. You lower the bird into the inverted cone; its head pokes out through the bottom, while its body stays snugly enclosed. Then you aim your weapon right to the raw edge of the cone …

Though bloody and distressing, a beheading can have miraculous consequences. The most famous headless chicken, an American cockerel by the name of Mike, was created by the slip of an axe. On September 10 1945, in Fruita, Colorado, as Farmer

Olsen took his weapon to Mike's neck, the blade missed the jugular. Although his head lay there on the farmyard floor, most of his brain stem and one ear was left intact; a clot formed, preventing him from bleeding to death.

Since most of a chicken's reflex actions are controlled by the brain stem, Mike was able to remain quite healthy, fed and watered with an eyedropper straight into his beak-free gullet. His intact voice-box also enabled him to keep crowing. Being an enterprising sort of a chap, Farmer Olsen took Mike on tour, introducing him to headless-chicken fanciers all over the Republic. Over the next 18 months, Mike visited New York, Atlantic City, Los Angeles, and San Diego where millions of patrons paid 25 cents to see him. He was valued at $10,000 and insured for the same.

Needless to say, over the more-than-sixty years since Mike's demise (on tour from choking), many have tried to emulate the miracle. So far, all they have got for their pains is a pile of corpses. Not that there haven't been loads of near-misses. Dianne who runs the local bakery remembers one well.

Growing up in my part of town in the 50s, Dianne's large family always had chickens. It was normal in those days – most people had a garden big enough to grow the family veg and have a shed for the birds. The children each ate two boiled eggs for breakfast every morning.

Dianne's dad was the one who looked after the chickens – he was often to be found outside in their pen, making sure they were clean and healthy; sorting out the cocks. They normally had four cocks – great big things, they were, and aggressive. Dianne thinks that's probably why the children weren't allowed in the pen. Whichever cock was fattest was the one they ate on Christmas Day. The kids had to do the plucking and the gutting; they never complained; they didn't know any different; they used to sit there on the concrete step outside the house, plucking away. And when they were done, they took the feathers to Dad for him to line the nesting boxes.

One Sunday morning, Dianne's big sister Carol was waiting on the step while Dad was in the chicken shed, decapitating dinner. Suddenly out rushed a bird, intact right up to its neck. Seven-year-old Carol turned tail and ran, with headless chicken in hot pursuit, flapping mightily. The girl ran screaming up the garden; once cornered at the end of the veg patch she rushed back towards the house. So did the gruesome beast. Up and down the garden they ran, Dianne and her siblings chuckling on the sidelines. It was particularly amusing, they thought, because Carol was usually such a goody-two-shoes; nothing had ever happened to her before. Eventually she could stand it no longer and took an almighty leap over the fence into Mrs Marchment's garden.

Dianne cannot remember how long it took her headless chicken to die – probably her father came out and gave it another going over with the axe. What she does know is that ever since then, Carol has suffered from an absolute terror of birds.

For me the thought of replicating such a horror was a major reason to forego the axe. Another was the fact that I didn't trust myself with a log, let alone a wriggling chicken in a traffic cone. Jay offered his services but I said no – it was my duty to see the whole thing through myself. I went off to research my method some more.

According to the charity Humane Slaughter Association, an axe is not a good idea because it too easily misses its mark. Also, after decapitation the brain may continue activity for up to 30 excruciating seconds. Poor Anne Boleyn.

So what were the alternatives? What about pliers? My friend Fred said if I didn't trust myself with an axe then try culling pliers, but the HSA website said absolutely not: the way pliers crush the neck is neither swift nor efficient, causing unacceptable pain.

What about neck-wringing – that classic way to dispatch a hen? Fred hated anything like neck-wringing; I think he felt it was too intimate a way to see them go. The HSA advised that for birds weighing less than 3kg (ie all the laying hens I had ever come across) careful neck-dislocation was definitely preferable to axes and pliers. Best of all, they said, was the electrical stunner – little tongs applied to the bird's head that zap their brains with a 130 volt charge, rendering them instantly unconscious.

You can purchase a stunner from a poultry supply shop. Remember to remove your bling, then set to work according to the instructions on the packet. Once the chicken is stunned, you need to cut off the blood flow to the brain as soon as possible. Get out your super-sharp knife and dig deep into her throat to sever both carotid arteries and both jugular veins. After a couple of minutes of blood-letting, you can be sure she will never run around your garden again.

I was happy with the idea of stunning my chicken. I have seen it on the TV – it looked a pretty humane way of doing things. My only problem was that the gadget would set me back around £700.

I gave the HSA a call. The young lady on the end of the line said if I really couldn't afford it (or it was an emergency), then I could go ahead and kill by hand. Not the way some barbarians do it, swinging the poor creature in circles around their head; just a simple click of the neck to dislocate the spine. She warned that this technique takes a certain amount of practice. She recommended I try it out on dead birds first. I didn't have any, but reckoned I could practice on my live one; I just had to remember to stop when I got to the point of no return.

I asked around, and found plenty of people who knew how to do chicken neck dislocation. If they were not currently a keeper, they had done it in the past, had learnt as a child or from a book now out of print. Many said the best time was first thing in the morning, before the bird had fed. It is common for it to defecate as it expires, and letting it feed up until the last minute makes this all the more likely. In my case, I could do it straight after the school run – by the time the children got home, the whole thing would be over. However, my friend John said when he was regularly culling, his preferred time was after dark, when the chicken was already half-dead with sleep. He said to offer only water for the whole day preceding, so the gut was as empty as possible.

As far as the technique went, it didn't seem to differ much from person to person. Tim had learnt from an organic poultry farmer friend whose flock failed to attain sufficient proportions to sell at the farmers' market. His expertise developed fast – 50 corpses in one afternoon. The most common problem, he felt, was that people let their emotions get the better of them – they weren't truly committed to going through with the job; didn't believe they were strong enough; just stopped when they felt resistance.

In the school playground, my friend Marie-Eve told me she had learnt dislocation and other methods as a child on the homestead in Canada. A crowd of children gathered round as their mums mimed out alternative chicken murders in intricate detail.

'When are you planning to do it?'

'This weekend.'

'Would you like a hand?' she asked.

'I feel I should do it myself.'

'Yes, yes. But I could be there – just for moral support.'

I thought for a moment. That might be a really good idea – a bit like a birth partner when you have a baby.

'Will you be my killing partner then?'

'I'd be honoured!'

Right. I can delay the moment no longer. My four-year-old happy hybrid is waiting. The time has come for her gentle dispatch, and you, dear reader, are welcome to join me. I am taking a deep breath.

By the way – don't believe the books that claim you can't learn to kill a chicken from a book; it's only because the authors are frightened that if they describe it, you are going to stop reading. Were you about to do exactly that? Then I suggest you try picking up the story again at the start of the next chapter.

OK. It is a cloudy winter evening, no stars to be seen; no frost nor wind; the garden is quiet except for the growl of distant traffic. The Ark's steeple stands black against the sodium glow of the sky.

Marie-Eve and I have knocked back half a bottle of wine, donned my muckiest allotment gear – wellies and macs, and left our husbands indoors with the children. We are searching around the garden for the best site – somewhere in the beam from the outdoor light, but not too close to the house in case things get noisy. If the dislocation doesn't work, we have agreed on Plan B – the axe. We also have a bucket standing by for blood, an old wooden pallet as a chopping block, and my sharpest kitchen knife.

I take hold of the handles of the Ark and gently ease out the side, expecting to discover my prey dozing on her perch. But no – she is snuggled up warm in the nesting box (take note – in preparation for her dispatch, I have wiped her name from my consciousness). As the cold air hits her she rises to welcome me; how trusting she is; she must be hungry. If it weren't for Marie-Eve hovering in the background, I think I might lose my nerve right now. Instead, I lift her big body under my arm and carry her to the centre of the lawn where the standby kit is posted.

With my right hand under her back, I tilt the body away from me and take a while to arrange the legs in my left hand (poultry keeper's fashion, fingers slotted between the ankles). It feels clumsy, but Marie-Eve is encouraging,

'That's good, that's good.'

I am familiar with the following series of manoeuvres, having been busy practicing them; the bird is familiar with them too and makes no objection.

I shift the position of my right hand round to the back, so the skull sits firm in my palm; it feels amazingly small and fragile. Of all things, the closest experience that comes to mind is of supporting a newborn baby.

Marie-Eve is crouching on the ground near the bird's head, checking on what I am doing.

'OK – I've got my first and second fingers slotted around the head,' I say.

'That's good, that's good.' These fingers will function in the same way as a hangman's noose. The rest of my hand is searching for the weakest point in the vertebrae, where the spine meets the skull. I think I have established where it is.

'I can feel her cheekbones.'

'That's it. Right up, tight against her face. Is your thumb on the top of her head?'

'Like that?'

'Good.'

I pull the legs upwards and the head down, stretching her neck as far as it seems comfortable to go. I turn the bird so her tummy is facing to the left, her back firm against my thigh.

On previous occasions, it was at this point that I released her and breathed a sigh of relief. But not now. Now I am focused on my new role as dispassionate killer. I lower my centre of gravity a little, stretching my toes inside my boots so my feet feel firm against the ground.

But the bird senses something. Suddenly this is not what she is used to. Her head strains upwards and the wings flap open – in the darkness they feel much bigger and more fearsome than I expected.

'It's OK, chicken – it's OK,' says Marie-Eve.

'It's all right, little one,' I croon. My daughters would be horrified that not only am I preparing to kill my pet, but that in her final moments I am telling her a fib.

The body relaxes and hangs vertically down again, as if reconciled to its fate.

I know the next move needs the greatest speed and determination to minimise any distress.

'I'm going to go for it.'

'You go.'

With my left hand pulling the legs up and my right hand around the head, I yank the body even longer (9 inches is what I am aiming for) and twist the head sharply back, against the curve of the spine. I know if I am too violent I will decapitate her, but decapitation is probably less painful than a wimpish strangling. I dig deep into the vertebrae, pushing the sections of bone against their natural direction. Can I hear a click of the bones dislocating? A gap at the base of the skull where they used to be connected? All I want is to know that the deed is done.

My hands are weak. The neck is strong; really strong. I yank and twist again – nothing. Now she is moving. Oh God – I haven't managed it. One more go – I grunt with the effort, pulling her legs awkwardly up in one direction, twisting the head down in the other. There is a crunching sound.

It that it?

The wings should be flapping. Flapping and flapping. The lady from the HSA said there is nothing conscious in the movement, just the nerves going AWOL. She said the bigger the convulsions the better – a sign that the body is well and truly disconnected from the brain. I had imagined holding the legs as this happened, her soul flying off, up and away from this winged body.

The wings *are* moving, but not out of control; more like an urgent objection to pain. She doesn't want to die. Argh!

'I think you've done it,' says Marie-Eve. I crouch down to her.

'I'm not sure – I'm really not sure …'

'OK – come here.'

I lay the writhing body against the wooden pallet and hold the feet fast; my friend takes the head.

'Wow – that head is throbbing.' The axe crunches.

'She's a tough old bird,' I say.

'God, she's tough,' says Marie-Eve, wielding the axe again – not high, just sharp and quick. As the wings flail, I grab the kitchen knife and sever the final artery. Released from Marie-Eve's hand, the head is twitching and the body is doing the same in exactly the same rhythm. How strange is this connection between being alive and being dead.

At last she is still. Probably it has only been 20 seconds since my first yank, but it felt like three times as much. Her comb is already turning pale and floppy; her eyes are tight closed. There is a small puddle of blood against the pale wood. I am glad it's dark and I can't see more.

Marie-Eve and I hug one another; we are shaking.

And then we hear Jay shouting from the back door.

'You done?'

Behind him huddle the children, dressed in fairy costumes from the dressing up box.

'Can we see, Mummy?'

As Marie-Eve and I scrub the pallet and sluice the lawn, Jay takes the kids down to the end of the garden and buries the head. Ellie gets them to hold hands and give Ruby a minute's silence.

'What did you do with the corpse?' asks Ellie the next day.

'It's hanging in the shed.'

'Ugh! What are you going to do with it?'

'Well, she wasn't ill; we shouldn't waste her.'

'You mean ...?' Ellie grimaces.

'You're going to eat her!' declares Beatrice.

'I'm going to make chicken soup. Will you have some?'

'I can't – I knew her!' cries Ellie.

'You don't *have* to, you know, Mummy,' says Bea, diplomatically.

'That's like eating one of us!' wails Ellie.

'It's like eating chicken.' I counter, just a bit smugly.

'But Ruby wasn't just any old chicken!'

'No – she was a happy chicken.' I say. 'All the other chickens we have ever eaten were miserable.'

When my Granny was little, her Sunday roast was probably a happy-go-lucky 'spring chicken' – a tender young male, plucked from the flock because he couldn't lay eggs. The other kind of chicken meat came from stewing a retired farmyard hen like Ruby to soften her stringy flesh. However, by the middle of the century, when Granny was writing 'Betty's chicken dish' into her recipe book, these by-products of egg-production were fast disappearing. The meat and the egg industries were becoming completely independent. And that's where it all got so miserable.

During the past half century, the efficiency of production in the chicken meat industry has risen a startling 450 per cent. This is mainly due to the selection of birds with a phenomenal ability to grow meat on their bones. These days, a meat bird is a completely different breed from a laying hen – a heavy, large-breasted creature of either gender (it never reaches sexual maturity), its body has been genetically refined to contain little real muscle; apparently we consumers have demanded this – so pale and clean is their flesh that it seems hardly like a dead animal at all.

According to DEFRA, more than 80 per cent of the chicken meat we buy in the UK comes from bog-standard broilers. Their short lives are spent largely immobile, stuffing their faces with high-protein feed. They grow incredibly fast, thereby avoiding the risk of too much disease and quickly making way for the next generation. Even before their adult plumage has fully developed they are big enough to eat – at five or six weeks of age. If you try boiling their carcasses, they quickly become a mushy gloop – that's how unformed the bones are.

From the shopper's point of view, such efficiency of production means our meat is cheap. It's the chicks who pay the price: so breast-heavy they can hardly stand, let alone skit about like normal birds, they endure crippling leg problems, heart failure and tumours.

On top of all this, the broiler parents (for, despite the industry's best efforts, chicken meat cannot yet be bred solely in a test tube, at least not without it costing a bomb) must suffer starvation. Think about it – cockerels and pullets only reach sexual maturity at around 18 weeks, by which time these breeds should be on the supermarket shelves. Without being starved, an 18-week broiler would grow into some sort of monster, so huge and unhealthy it would be unable to produce eggs or sperm (the breeders have methods to save them the bother of copulation).

And just in case, as consumers, we thought spending more money would solve things – even the more expensive birds aren't necessarily having a brilliant time. A 'free-range' label (2 per cent of the market) means they ate the same 'growers' feed' as cheap broilers but had access to outdoors; their minimum slaughter age was eight weeks rather than five, so it is possible they were a slower-growing strain. But not necessarily – they might have been fast-growing but made to slow down, again through starvation. And even if they were given access to a run, a fast-growing breed probably didn't have the time or energy to make use of it.

Our best bet is definitely the 'organic' label. It indicates that the dead chick was fed the most sustainable kind of feed, kept free range and slaughtered at as old as 12 weeks. Ideally, it was one of the slower-growing breeds, with a bit of free time to walk around, and able to make it to three months without having a heart attack. Because of its diet and the length of time it lived, organic chicken is the most expensive, and therefore forms less than 0.5 per cent of the UK market.

And the market is vast. According to the RSPCA, in 2008 830 million broilers were slaughtered in the UK (roughly 14 birds per man, woman and child). Henocide. Of course, that figure excludes all the possibly-more-miserable broilers we imported, stuffing their meat into sandwiches and pies before anyone asked where it came from.

I feel wretched about Ruby's death but I feel even more wretched about the death of all the broilers I have eaten in my life. Merciful but not kind, it entails shackling them upside down and sending them struggling along a conveyor belt into a water bath where, with luck, they are electrocuted to death. I say 'with luck' because if that frightened creature happens to raise its head as it travels towards the bath (as Ruby raised hers as I approached the point of no return), it may well not receive the full voltage.

And what's worse than a dead broiler? A half-dead broiler.

Ruby's carcass hung in the garden shed for nearly 24 hours, the blood dripping into a bucket. I hoped the process might tenderise her meat; I also hoped that next time I held her she would be well and truly cold. The night before, I had found the warmth of her headless body really disturbing; no different from holding her when alive. Instead of plucking her while warm, as the old poultry books recommend, Marie-Eve and I hoisted her by the ankles and went inside to finish off the wine.

Next day, the body was reassuringly stiff as I filled the compost bin with her feathers. It was good she didn't have a head – it made her more anonymous. Plucking the body was not difficult; the tail took a bit more effort, but the wing feathers stuck hard. Jay took his penknife to them, and then set to work chopping off the feet and carefully prising out the crop (strangely full, after all my efforts at starving her). At the tail end, he made an incision between vent and breastbone, thrust thumb and forefinger inside and drew out streams of gut and reproductive tract. At this point I summoned our daughters for a live anatomy lesson. When the first bright yolk appeared, full-sized with a filigree of blood vessels across it, we were all amazed; then two more yolks descending in size, the second with a very evident germ cell sprouting from its surface. Behind these came a string of baby-eggs, at least a hundred of them, getting smaller and smaller along the stem of the ovary. Despite her slowing in productivity, her whole body cavity was still stuffed with potential.

After a good wash under the kitchen tap, Ruby lay on the chopping board like any other healthy piece of meat. I had expected her layer's body to be scrawny and was pleasantly surprised how very chicken-like it looked. I had anticipated the flesh being dark, like a pheasant's, but in fact the breast was quite pale and soft to the touch. The legs had good muscle-tone from all that running around.

Jay and I looked forward to an excellent meal.

One thing I should add in anticipation of eating our bird is that Jay is Jewish – not practising, you understand, but the bearer of plenty of cultural baggage, especially when it comes to Jewish Penicillin. His mother used to make soup when he was sick; it was always delicious and sometimes curative. He remembers the recipe exactly – the carcasses carefully frozen over many months; the juices from several roasts likewise ... I should know it off by heart, for each time he is ill he reminds me, but somehow I always manage to blank it out. What I can offer instead is what happened to Ruby:

Ruby Soup

Put the whole chicken in the pressure cooker with about a litre and a half of water, along with carrots, celery sticks, a large onion (peeled and quartered), half a head of garlic (peeled), half a lemon, a couple of bay leaves, some stalks of parsley and coriander and three slices of fresh ginger.

Bring to boiling pressure, then leave to simmer for 20 minutes.

Open up the pot and fish out the solid stuff with a slotted spoon. Return the garlic and carrots to the stock, along with meat from one leg and a little breast. You can reserve the rest of the meat for another meal – mixed with home-made mayonnaise there's no better sandwich filling on earth.

Liquidise the soup. Add more lemon juice and salt and pepper to taste.

Jewish people have a tradition of poultry sacrifice called *Kapparot.* Still common among the ultra-Orthodox, especially Chasidic communities, it takes place during the autumn when, conveniently, chickens are likely to be coming to the end of their laying season. The idea is that every man in the family should sacrifice a cock and every woman a hen; pregnant women get one of each. If you are seriously into the ritual, you lift said fowl above your head and revolve it three times. Meanwhile you recite the following:

'This is my exchange, my substitute, my atonement; this hen/cock shall go to its death, but I shall go to a good, long life, and to peace.'

The bird is then slaughtered and given to the poor, or you can cook it according to one of the myriad of Jewish chicken soup recipes and share it in your community as part of the feast before Yom Kippur.

Some people object to the *Kapparot* ritual because the notion that a dead chicken frees us from sin is just too simplistic. Others say that the sacrifice is just the beginning of the process of atonement, inspiring you to *teshuvah* – a period of repentance in which God weighs up your sins. What I like about it is the acknowledgement that a hen and a woman (or a cock and a man – *gever* and *gever*) are interlinked, that the destruction of one arouses the destruction of something in the other. I think it fits nicely with that whole Jewish Penicillin thing.

According to my research, the University of Nebraska Medical Centre has *proved* once and for all that chicken soup has medicinal properties. They think it is something to do with the way a fragrant broth inhibits the clumping of white blood cells (neutrophils) that cause congestion and inflammation when you have a cold. If you have ever had a bowl of golden, clear and aromatic broth while you have been ill, you may well confer. Or you may prefer more metaphorical explanations; you might well feel that chicken soup is eaten not just with the digestive and olfactory systems, but also with the emotions.

Which in my family's case meant our twenty-first century children making a right fuss over supper. Ellie swore that from now on she was officially vegetarian, that eating animals was as bad as eating one's parents. Bea bravely nibbled some breast meat and declared it 'just like real chicken!' Jay and I agreed that the soup was far superior to any we had ever tasted, its rich flavour augmented by our sense of satisfaction at having grown it ourselves. Each mouthful was sweetened by thoughts of all those eggs Ruby had given us (almost £300's worth, we calculated), plus nearly four years of garden company, from POL to pot.

And even if my daughters never get to do *Kapparot*, even if (God forbid) they never learn my superlative method of cooking chicken soup, they may still recall the sacrifice we made with Ruby.

'Gross!' said Ellie.

But still she sat there, sharing our contemplation, if not our meal. Helping us atone for all those miserable broilers we had eaten so thoughtlessly in the past.

Birth Plans

When the rooster starts to crow
Grab your partner on the floor!
Slim Gaillard

As the late great gardening writer Christopher Lloyd used to say – every death is an opportunity. What my Ruby soup had done was to whet my appetite for more. For more soup, even. Having undergone such an important rite of passage, the challenge was on to improve my dislocation skills. If I could make friends with someone like Tim's poultry farmer, I could get in some serious practice before I had to sacrifice my own bird again. I felt quite excited at the prospect of killing more happy birds and thereby permanently shunning the evils of the broiler industry. Perhaps I could persuade the allotment membership to let me use a shady communal corner that was rapidly turning into a dump. Then I could keep a decent sized flock.

The Chair of the committee was out on his plot, putting his dahlias to bed for the winter. When I mentioned my plan he frowned.

'Oh dear. I don't think the other plot holders are going to like that.'

'Why not?'

'Well … what if they get out?' His face looked strangely animated at the thought.

'That shouldn't happen …'

'But if it did – they could really do some damage, couldn't they!'

'More likely, the fox would get them.'

That should have drawn a line under the matter. But no; the Chair was on a roll; perhaps he was an alektorophobe.

'And what about avian flu?'

'The UK has been officially free of it for nearly four years …'

'But what if it comes back?'

'Then I would do whatever DEFRA tells me – there's loads of advice already on their website.'

I wanted to say – what about the ducks? What are you going to do about all the ducks and geese from the river, flying over the allotments? But I resisted.

'Well, I'd talk to the council if I were you,' he concluded.

When I phoned the council's Parks and Leisure department, they said,

'Look at the deeds'.

I said, 'I have and there's nothing about chickens in there.'

The man on the end of the line sounded stressed.

'The site is owned by us?'

I said, 'That's why I'm phoning.'

He said, 'Hang on a moment.' After a good few minutes he asked me to phone back later. I phoned back later. He'd gone home. I phoned back next day, feeling a bit of a sucker for punishment.

'Hi,' I chirruped, trying to sound as though we were old mates. 'I was just wondering if you'd found out anything about the chickens?'

'Oh yes,' he said, shuffling papers. 'Here we are – it says 'No Livestock on City Land.'

'Are chickens livestock?' I enquired, in as mild-mannered a tone as I could muster.

'I'm afraid they are regarded as a ...'

'... a health and safety risk,' I said.

Well, at least I'd stolen his punchline.

In fact, the conversation did not end there. We went on to have a long and animated discussion about food production and allotments. I tried to argue that unused corners should be turned over to useful food production; he suggested they were better left for wildlife. I explained about the wretched meat industry and how we all needed to find ways to improve animal welfare ... I said the council should change its rules; he said I could write in and make my suggestions, but he couldn't promise a prompt response, Parks and Leisure was an overstretched department ...

Eventually I gave up. I was angry and upset. But a friend of mine was determined I should not be so easily defeated. She offered to get Googling, and soon enough she had found what she was looking for.

It turns out that if you get the idea of keeping chickens on your allotment, rest assured – according to the Allotments Act 1950 Section 12, you have a statutory right to do so; the consent of the appropriate municipal authority is not required and need not be sought. The Act advises that you 'erect or place and maintain such buildings or structures on the land as are reasonably necessary for that purpose.' The same goes for rabbits.

*

So this was my plan: come the spring, I would erect an aviary in the shady corner and adapt the children's old Wendy house as a cheap coop. Or if the committee couldn't bear the idea, then I would get a couple of Arks from Solway Recycling and offer my fellow allotment holders a free weeding and manuring service.

As far as the breed was concerned, I felt ready to move on from the commercial hybrids. I wanted something good for meat as well as eggs with not too wild a personality: what used to be termed the 'utility' or 'dual-purpose' bird. When I was little there were lots of kinds that fitted the bill, the best known being my grandfather's Rhode Island Reds.

Named after the place it was born – the Little Compton district of Rhode Island, the RIR was developed towards the end of the nineteenth century, originally a cross between the Brown Leghorn from Italy and the sturdy Shanghai – the best layer and the best meat bird. Pioneers of large-scale food production, the Americans understood that the most economic breed was going to be good to eat as well as good for eggs. It was logical enough – instead of throwing the boy-chicks in the grinder, why not bring them up to deliver a decent Sunday lunch?

In the days before mass production, RIRs became popular with people like Grandpa because of their strength, their reliable reproductive abilities and the determination of their publicists. With numerous supplementary breeds introduced into the stock, they arrived in the UK in 1903 and were standardised in 1904. These days, Rhode Island hens are a common ingredient for commercial hybrid layers; for those of us outside the commercial sector, there are exhibition strains with beautiful glossy, chocolate-coloured plumage. The trouble with these is that they are no longer 'utility' at all.

I wish I could have continued Grandpa's legacy, but I needed to be practical. If RIRs could no longer produce meat for the table, I should consider something else. My friend Karen suggested her still-chunky American breed, the Wyandotte. Named after a Native American tribe, this utility bird was developed by crossing pretty crested Polands and Hamburghs with the stately Brahma. Though Karen admitted that her cock was proving too aggressive to stay out of the pot much longer, I found his smooth-fitting partridge plumage and athletic body strikingly attractive. He wore his comb tight to his head like chain-mail, instead of high and spiked. The range of colours currently standardised in the UK totals 14, so I should be able to find something to match the Chair's dahlias. My only potential problem was that the Wyandotte's athleticism gave it excellent flying potential which meant it really might escape on to other people's plots, unless ruthlessly wing-clipped. It is also happiest with plenty of space, which I definitely wouldn't be able to offer with an Ark.

Regarding these limitations, I turned to one last useful American – the Plymouth Rock. This was the utility bird people used to keep in the UK before the hybrids took over; quite possibly it was the one Dianne's dad had in his garden up the road from mine. Developed in New England in the 1800s and registered in 1869, it is still often the basis for broilers around the world. Happy enough in a confined space, yellow skinned and tenderhearted, it would be perfect for keeping in a decent-sized Ark. It is also rather pretty, with lots of colours to choose from, though the black and white striped (known as 'barred') is most common.

My only problem was that it seemed impossible to buy a Plymouth Rock (or either of the other two, for that matter) for under £20. I knew that traditional breeds were not going to be as cheap as hybrids, but I had rather hoped that keeping them for meat would prove economic. I was going to be spending quite a sum fattening them up before they were ready for my dislocation skills. And all the time, in the back of my head, I wouldn't be able to forget that their cousins' corpses at the farmers' market cost under a tenner.

Central to the problem of money was the problem of egg production or rather, lack of. During the 1950s and 60s, as the hybrid industry was taking off, there were very organised laying trials across the UK to establish which traditional flocks were the ones to breed from. Doubtless the utility strains were in there. But gradually, as the commercial hybrids raced ahead down their own production line, breeders of the old strains lost interest in keeping such records. These days there is no quality control, no recording of the productivity of parents or offspring.

Sadly, traditional birds lay much less than they used to. While the Victorians could expect a Plymouth Rock to produce around 285 eggs a year, we have to make do with 200. A contemporary Wyandotte might offer a similar number, but probably less. And what about the RIRs? Grandpa had expected his to lay at least 250. From the websites, I could see that some in the US still boasted the same, but in the UK they seemed to have gone right down. To 150.

In my financial calculations I also had to factor in the mortality rate, though I hoped not the shocking commercial broiler statistic of 1 per cent a week. Down on the allotment I was in danger of losing more than I would in the garden. There would be less human activity, so more foxes; new predators in the form of badgers and perhaps birds of prey. Then there would be the hen-nappers. Though the site has a padlock on the gate and I'd have another on the pen, I knew what happened to another flock in the area – the local lads decided torturing a bunch of birds was good for a laugh.

One way to keep down the price would have been to get them as one-day-old chicks and fatten them up at home; cheaper still would be to breed them myself. What I could do was buy just a couple of Plymouth Rock pullets from the best stock, get in a cock and breed a whole new flock for my meat 'n' egg plot.

My friend Sally, who has bred chickens all her life, warned me I was embarking on a whole new enterprise here. I reassured her that I would avoid keeping a cock because I didn't want run-ins with my neighbours; I left that to my friends. I suggested borrowing one of her Three Teds but Sally said there was no use in that – their progeny would fail both my meat and egg targets. If I asked around amongst my poultry-keeping friends, someone was bound to know of a good cock to rent or borrow; or else I could get in touch with him via his club, once I had decided on the breed.

Right. I reminded myself that this was my first foray into cross-breeding; I mustn't be too ambitious. For example, I mustn't borrow a cock belonging to one of the aggressive breeds and risk him harassing the kids. The point about cross-breeding is to introduce new qualities – so, if the mum's breed was a Plymouth Rock and good for meat, then Dad might as well bring some laying abilities with him. Though breeders say egg laying capacity is not a matter of breed but of selection, it was worthwhile confining my searches to a breed that had not lost all its talents in that area.

I would love to have tried an Italian Leghorn, with its stylish looks and excellent egg laying reputation. My problem was its extrovert Mediterranean nature: the sound of the cock's crow would reach the other side of town, and his enthusiasm for roosting in trees struck me as foolhardy, regarding our local fox population.

My first choice would have to be the much milder-mannered Great British utility bird – the Light Sussex. This breed has a nice backstory: a century ago, the farmers of Lewes got together and realised they had only one flock of genuine Sussex hens left. Having formed a poultry club, they rescued the local breed from extinction and gradually established a range of colours, the light one always proving the most popular. Its striking features – a white body with a zig-zag black collar and black tail, mean that in the past half century the Light Sussex has become popular as an exhibition supermodel and in the process has lost much of its laying power. Fortunately, my friend John informed me, long ago a few flocks were exported across the Channel where French farmers preserved their usefulness. Someone in John's local poultry club might smuggle one back to Blighty for us.

In the meantime, I could go for the original French breed that John himself used to keep. The Maran has grace and poise, and can become very tame if handled frequently. Famed for its mahogany-tinted eggs, it was developed as long ago as the 1800s in Western France by crossing the Asiatic Langshan with the dominant gaming birds of the region. The Marans standardised in Britain are clean-legged strains, Dark and Silver Cuckoo; those bred from the original French stock have feathered legs – the most popular being Copper Black with its petrol-green plumage. If I could find a strain that had retained its laying abilities, this might well turn out to be the one for me.

*

John reminded me that introducing a cock could be tricky, what with the hens' uncompromising pecking order and all. Having only two potential mothers to choose from should be easier: at least there wouldn't be others left behind, feeling jealous. Penning birds in a *ménage à trois* is common practice in chicken circles, and even has its own, socially acceptable name – a trio.

Though ideally I would have put my trio on the allotment, the Allotments Act 1950 Section 12 said I couldn't have cocks there, so they would have to do their thing in the garden. A cock generally has a repertoire of tricks with which to impress, and my children were looking forward to the show. They wanted to see him strut his stuff, his large comb and long tail extended to their fullest potential.

We would recognise a less confident seducer if he resorted to using the call that is meant to draw attention to a tasty morsel of food. The more honest approach would be the John Travolta style of seduction: standing sideways on, he tilts his body in towards the female, showing her the beautiful plumage along his back and tail. Then he lowers his gorgeous wing to the ground and begins to circle her, to the accompaniment of his own, excited squawks. The downward thrust of his body is almost supplicatory, as if on bended knee. Round and round he staggers, round and round. If she shows no interest, he doesn't give up but turns and performs the same steps with the other wing lowered in the other direction. Perhaps she will fancy him better from this side.

Eventually, with any luck, he triggers a response: the hen cowers to the ground, making it a cinch for him to tread her.

The way a cock mates a hen is called treading rather than mounting because he literally treads on top of her. Ouch. Worse still, while his spurs are scrabbling on her back, he keeps his balance by holding on to her neck feathers with his beak.

Grab your partner on the floor!

A really randy cock treads as many times a day as he can. Of the two hens I would be offering him, he might prefer one to the other and end up shagging her featherless, poor dear. John said in that case give her the day off and make him make do with monogamy.

My neighbours would be relieved to hear that the cock did not need to be around for long. After about a week we could be pretty certain the eggs were fertile. The sperm then hung about inside the hen's body for a couple more weeks, so even when he had departed, we would still be collecting fertile eggs. For hatching, I needed to choose the ones that were perfectly smooth and clean.

Contrary to popular belief, fertilised eggs do not need to be kept warm, in fact they should stay cool to stop them developing prematurely – around 12 degrees Celsius is best, stored in an egg box (with their pointed end upwards) in a shed or larder. They can wait for a couple of weeks, if necessary – just as they would were she accumulating a clutch of her own.

Then I needed to decide what to do with them. The trouble with hens is that presented with a clutch of fertilised eggs, they don't necessarily want to sit on them. With commercial hybrids, that's because the nurturing genes have been bred out of them. Not surprising when you realise that all through the incubation and the first few weeks of the chicks' lives, the mother will lay not one single egg. Broodiness is absolutely the last thing a poultry farmer needs. Like many in the non-chicken world (in my experience, boyfriends), he has it down as a serious affliction.

The signs are obvious – she starts going gooey over other people's babies. Nearly. She refuses to come out of the nesting box; when the poultry farmer tries to pull her out, she attacks him; her breast is puffed up and feverish to the touch. In *Fowls and How to Keep Them,* Mannering says if you want to cool her ardour then plonk her in the run or some other airy place, or else pen her with 'a vigorous spare cockerel', though sometimes the broody hormones are too torrid even for that. John advises using a box with mesh on the floor, a bit like a battery cage – the draughts put her right off nesting.

Of course, my problem was going to be quite the opposite – how to get the hormones going. What if my hens waited until their fertilised eggs were addled before getting broody? Just because I now realised that the meaning of life lay in the next generation, it didn't mean my hens would. They might decide they preferred successful careers in egg production.

Which is why incubators were invented.

I have vivid memories of eggs in an incubator when I was at primary school – taking it in turns to press our faces against the murky window and check for signs of life. My daughter Beatrice's class recently borrowed the same system from an agricultural college.

I don't remember if my childhood incubator had an automatic egg-turning mechanism; certainly the contemporary one did – it saves a lot of bother as the eggs need turning at least three times a day for the first 18 days. The turning has to be in both directions, to prevent the yolk cords from twisting too tightly, and an odd number each day, so the eggs don't lie on the same side on consecutive nights. I don't remember adding water to the trays to increase humidity or opening and closing the air vents to control airflow, duties which Beatrice's teacher accomplished with utmost dedication.

I do remember 'candling' the eggs at the end of the first week – closing the classroom curtains and holding each one up against a bare light bulb to see whether a network of veins was beginning to form. Then the same a week later, checking for that little bean-shaped blob of an embryo. Seminal experiences these – almost as heart-stopping as thirty years later, lying in the radiographer's darkened room with my own foetus inside me.

The eggs need to be of similar size so they all incubate at the same rate. It doesn't matter if they have been collected over a period of a week or more – the important thing is that they all go into the box at the same time; that way they are most likely to hatch together. Come the time for hatching, the full-grown chicks manage to communicate with one another through their shells using a tiny clicking noise, as speedy as a vibration –

'Clickety click – fancy taking the plunge?'
'Click, clack, click – I'm up for it if you are.'
'Click, click – see you out there, guys.'

In an incubator this event is a charming little miracle, but in nature it can mean the difference between life and death. Because Junglefowl and their relations nest on the ground, their young are very vulnerable to predators; at least if they synchronise their birth they can abandon the nesting site together and have the best chance of survival.

Depending on the efficacy of antenatal communication, the hatching period may last as long as 24 hours. In the old days, this was the time we kids were jostling for a chance to look through the window. These days, the teacher's dexterity with her camera meant that Beatrice and 29 other five year olds had no such problems. In glorious Technicolor up on their whiteboard, they watched as one chick used its tooth (that's its one, special 'egg tooth', on the top of its beak) as a little pickaxe. A muscle called the pipping muscle on the back of its neck gave it the extra strength to hack its way out of the shell.

The birth accomplished, the children were desperate to hold the new baby. But the teacher forbade their opening the incubator until all its brothers and sisters had arrived. Like opening the oven when a soufflé is in there, you need to be sure your eggs are all fully cooked or else you risk serious disappointment.

At the end of the afternoon, their down dried and chirping noisily, the babies were moved into their 'brooder'. The school brooder (then and now) was a large cardboard box lined with wood shavings, sitting on the carpet with an infrared heat lamp hanging over it. That lamp keeps them at a constantly cosy 35 degrees Celcius and gives them a lovely pinkish glow. It is on a stand that you can raise every week in order to gradually lower the heat until the chicks are fully feathered and able to chirp at regular room temperature (any time between four and eight weeks).

Chicks are the most darling creatures to observe, tottering about, so full of determination and yet so vulnerable. When you pick them up they are light as a dandelion head, their bodies sending a tiny electric whirr into your hand. They are born with all sorts of communal instincts, like the urge to follow one another, to huddle up if they are cold or to spread out if they are too warm. Beatrice's class knelt entranced at the sides of the brooder; it was all their teacher could do not to climb in and give the cuties a cuddle.

One young chap stayed at the incubator, looking down through its window.

'Miss?' he called, 'What about these ones?'

'What about them, Sidney?' croaked the teacher, trying to remain unruffled.

'Are they dead then?'

'I'm afraid they probably are,' she said. And he gazed even harder through the glass as if therein lay some answer to the mystery.

I resisted the temptation to interrupt Sidney's musings with the line that had hopped into my head – '*Don't count your chickens before they hatch!*'

It originates in Aesop's fables – the one about the milkmaid and her pail. A young girl on her way to market carrying a pail of milk on her head, plans how she will sell it and buy some chickens. The chickens will lay eggs that she will sell and with the money she will buy a beautiful dress and hat. At the thought of displaying her lovely garb, she tosses her head and the pail falls to the ground. When she gets home, her mother admonishes her with that now-famous proverb.

The odd thing about it is that, at the time of writing, chickens had yet to be introduced into Greece; the Greco-Persian wars being still a good half century away. It therefore seems unlikely that Aesop chose the metaphor. I suspect the chicken bits were added ages later. Which begs the question, what was that milkmaid really planning?

Talking of planning – none of Beatrice's class enquired what was planned for the fluffies when they returned to agricultural college. Neither the teacher nor I dared spoil things for them. If Sally had been there, I think she might have done. Her Three

Teds were a reminder that unwanted boy chicks hatch out just as often as girls. Sally had planned to cull hers when they were born but discovered she was one of those people who loses their killing ability as they get older. My problem was that I wouldn't know how to identify them.

Sally said it certainly wasn't easy. Probably the first I would know would be at around 10 or 12 weeks, when the males started to crow. A little earlier than this, if I was looking carefully, I might see the cockerels grow their combs sooner than the pullets, and have thicker thighs. Some traditional breeds have differing feather growth – in heavier breeds, the girls get grown-up plumage earlier; in lighter breeds it is the boys who grow their tails first; in some hybrids the young females have longer wing pin feathers than the males … Which is all very well if you are experienced with whichever breed it is, but well-nigh useless for the rest of us.

Unless we know how to sex day-old chicks. It is called 'vent sexing'; this is how it's done:

The sexual organs of both male and female are hidden away in their vent, tiny because they are so young. Picking up that fluffy ball, the vent sexer squeezes its back end and takes a good look inside. He or she has studied the characteristics of chick bits in minute detail, committing to memory the fifteen basic patterns that manifest themselves in different breeds. Broadly speaking, whatever breed s/he is inspecting, s/he is looking for a bump that indicates the chick is male. Some females have very small bumps, but rarely do they have the large bumps of male chicks.

This highly specialised craft originated in Japan. During the 1920s, Professors Masui and Hashimoto from Tokyo started giving lectures about it, but no one took much notice. Then in 1933 they wrote a book and an English translation was published in Canada under the title *Sexing Baby Chicks.* Suddenly, poultry breeders throughout the US and Europe became interested, employing the protégés of Masui and Hashimoto on a seasonal basis via the Japanese Chick Sexing Association. At the Kibworth Hatchery in Leicestershire, a breeder called WP Blount learnt the technique from his Japanese employees and published his version of their book called *Sexing Day-Old Chicks*. In it he confessed that he was neither as dexterous nor as reliable as his teachers, who could sex 3,000 chicks a day with 99 per cent accuracy.

In 1941, after Pearl Harbor, seven of the seasonal vent sexers got stuck in Great Britain and interned on the Isle of Man. Having completed a particularly successful sexing season, they were flush and found that their cash made them the most popular internees in the camp. When they were eventually released, three of them decided to stay on permanently in the UK, passing on their specialism to those who wished to undertake the training.

Fortunately, thinking again about Bea's chicks, I realise now that they didn't need vent sexing. They fell clearly into two distinct kinds of plumage – yellow and tawny. Most likely the colour was directly linked to their gender, either because they were sex-linked hybrids like the ISA Brown, or because they came from an auto-sexing breed.

Autosexing was invented during the World War I by an English professor called Reginald Crundall Punnett. When he crossed a barred (stripy) Plymouth Rock with a brown Campine he found the boy chicks came out pale yellow and their sisters came out barred and buff coloured.

These days there are lots of autosexing breeds available – you can tell because in their name they have the suffix 'bar' – for example, 'Legbar' from an original Leghorn cross, and 'Brussbar' from a Sussex. It is extremely convenient for the breeder. Where sex-linking required parents that came from particular, separate breeds, the autosexing characteristic is there within the one breed and continues down the generations.

Their handy colour-coding proved a Godsend for a friend-of-a-friend when the eggs she had been incubating failed to hatch.

It was a very special clutch of eggs, bought from a reputable breeder – the first chicks the family would ever own. Mum had promised the children that they would watch the babies emerging from their shells, that they could each give one a name and welcome them as adorable new members of the family … She worried what exactly she would say if any of the chicks turned out to be male. She knew there would be a dreadful scene if suddenly at 10 or 12 weeks or whatever age their testosterone revealed itself, the pets had to be confiscated. She decided to put her worry on the back-burner.

Every morning, the children leapt out of bed and rushed downstairs to check the incubator. Day 20 went by, then Day 21, then Day 22. The children grew anxious. The friend-of-a-friend grew desperate – how was she going to confess to her kids that all these weeks of waiting and watching had come to nothing? That all they had was a clutch of rotten eggs. Day 23 – after a sleepless night, she phoned the breeder and together they hatched not a chick but a plot: after she had dropped the children at school she would zip over to the farm and pick up five brand new buff-coloured chicks, born that very morn. And the marvellous thing was – they were all, incontrovertibly, females.

*

Back in Bea's classroom, the first thing the children did once the chicks were settled under their lamp was to offer them something to eat and drink.

Water is safest delivered in a chick-sized water fount, the grown-up ones being dangerously large for a chick who has yet not learnt to swim.

The food they most appreciate is chopped hard-boiled egg. As everyone who has ever eaten breakfast knows, hard-boiled tastes totally different from raw, so we don't have to worry that this early experience might cause them to eat their own eggs later in life. Finely chopped chickweed and even grass get gobbled up, as do dandelions and clover.

Bea's chicks were given 'chick crumb' for their very first meal, mixed with a little warm water so it was swollen and porridgy. Gradually the dry version was introduced over the following days, and that became their staple for the next few weeks. Chick crumb is a high-protein feed containing medication to keep coccidiosis at bay.

This killer disease occurs mostly in chicks around three to six weeks old, often in warm, damp conditions. It is caused by coccidia protozoa (single-celled parasites) that breed in the chickens' intestine and spread through faeces. Large-scale breeders use spray vaccines to guard against it, but there is no such facility for us smallholders.

If the chick picks up some coccidia on its foraging rounds, it can get sick very quickly indeed, hunched and cheeping miserably in the corner. A well-known sign of infection is bloody droppings, but unfortunately not all forms of the disease produce this symptom, and anyway if you see blood the chick is probably too far gone to be treated. The parasites multiply speedily inside its body, so it needs to be removed from its fellows and taken to the vet FAST.

Treatments come in a bottle and are administered with a pipette into the crop.

Infected birds excrete the coccidia eggs in droppings, thereby contaminating their bedding and the ground for up to two years. If you think you have a sick chick, you need to change the bedding every few hours. For the ground, normal disinfectants do not kill the eggs, but ammonia does, as well as temperatures above 56 degrees centigrade. A blast from the midsummer sun, or a few days' dousing with boiling water and household ammonia should do the trick.

With any luck the chicks develop immunity to coccidiosis as they get older.

By the time they are six weeks old they are ready to be weaned on to something called 'growers feed' – lower in protein than the chick crumb and without the medication. They also need grit to get their gizzards working, especially if you are starting to introduce grain. This new diet can be mixed with the chick crumb for a few days so they don't get tummy aches from too rapid a change. Growers feed comes in both crumb and mini-pellet form. Layers crumb or pellets come on the menu when they reach POL-stage, around 16–18 weeks old.

Some people have hard-and-fast objections to starting their babies on such a path. As someone who tries to avoid industrial feed, I do sympathise. But it makes life much more expensive when you decide to feed your chicks on pin-head oatmeal and millet. It also makes life harder – for a few months, you are watching like a hawk for the tiniest signs of sickness. And because coccidiosis thrives in humid conditions, without the medication you must be absolutely fastidious about keeping the chicks' bedding dry and changing it regularly. Leading to the old rhyme – *'Dirty and dry, not much will die. Dirty and wet makes cash for the vet.'*

Not that my friend Dianne from the bakery remembers her chicks ever needing to go to the vet. Half a century ago, when she was growing up in my area, the family had a constant stream of pretty chicks. And it was her job to look after them.

Every Thursday her mother bundled together some old clothes and set them by the front door. Dianne remembers it was Thursdays because that was baking day – a warm, yeasty smell filling the house. When they heard the van hooting its horn outside, the kids would pick up their bundle and run out into the street to hand it over. A friendly chap he was, Mr White the rag 'n' bone man, with his soft white hair and such a gentle manner. In return for their clothes, he opened up the back of the van, revealing two big boxes: one was a tank of goldfish and the other a flock of little yellow chicks. The children were meant to choose between them, but in truth they weren't allowed a fish because it would have been useless. Anyway, Dianne always loved to take away a fluffy, chirruping ball in her hands.

They lived in the back room in a cardboard box, feeding on kitchen scraps boiled to a mush on the stove. One time she remembers Mr White had some sort of job-lot to get rid of and her mother agreed to take 100 at once. They had to let them run free in the room, there were so many of them, scuttling between the legs of the benches and the dining table. It was quite a sight, all those yellow pom-poms darting about the floor. Of course, they put newspaper down for their droppings, but they were so tiny it was no bother. Dianne thought it was a great treat, having so many babies to care for, and felt sad when their adult plumage arrived and they had to go outside to Dad's shed.

I fancied turning over my back room to a flock of chicks for a while. I reckoned I would probably give them the medicated crumb, but they would still be vulnerable to all the diseases against which hybrids are vaccinated. Marek's is especially bothersome – formally known as Fowl Paralysis, this herpes virus affects the nervous system, causing limb paralysis and blindness; if it doesn't kill, it permanently incapacitates. Though vaccinated, one of my hybrids might have carried Marek's and left traces around the garden. That meant extra special care and fingers crossed when the chicks went outside. An outdoor run would be good for when they were tiny (vulnerable to rats and even to our ancient cat) – I would leave it in sheltered spots and move it around regularly, giving them constant access to what I hoped was fresh, clean grass.

When they were a bit bigger, I looked forward to offering them my whole garden to hone their foraging skills. Unlike full-grown hens, chicks can do little harm to the lawn or the flowerbeds. New mother as I was going to be, I intended to be more fastidious than ever: I would make sure the ground was not wet – wet down can cause a nasty chill; I would mow the grass before they went on it to prevent my babies getting impacted crops or accidentally choking; I would watch them like a (vegetarian) hawk.

And what about the Allotment Act 1950's insistence that the cockerels stayed home (once I knew which ones they were)? Utility breeds are much slower growing than any of those commercial broiler types; the males might not be ready to eat before they were seven months old. That meant a lot of crowing for my neighbours to put up with. If I was lucky, they would take a long summer holiday; followed by an autumn of apologies. Come Christmas, I would try and make up for it by inviting them round to share my supremely delicious, home-grown coq au vin.

Brit Chicks

Oh! De Shanghai!
Don't bet your money on de Shanghai
Take a little chicken in de middle of de ring
But don't bet your money on de Shanghai
Stephen C Foster

When Julius Caesar and his army invaded Britain, they were surprised to discover that *Gallus gallus domesticus* had preceded them, probably in the company of the Phoenicians. What surprised them all the more was that the birds were kept not for food but for fighting.

Though the Brits did gradually come round to the idea of eating their chickens' meat and eggs, it took several hundred years for them to relinquish their taste for cockfighting. Between 1833 and 1845, various acts of Parliament were passed that eventually banned it. And having at last acknowledged that birds tearing one another apart was unacceptable entertainment, the Victorians discovered the joys of breeding them. As trade routes and transport systems opened up, amateur and professional breeders began importing all sorts of interesting new breeds to replace their fighting stock. The most important being big, docile birds from the East.

In 1842, when the port of Shanghai opened to international trade, ships were launched carrying specimens of exotic Chinese chickens previously unknown in Europe or the Americas. The biggest and most docile of these was imaginatively christened the Shanghai. She went first to the US (where Stephen C. Foster came across her), but when a Mr Burnham of Boston judged her fine enough for an Empress, she was soon on her way across the Atlantic.

Renamed Cochin on arrival in England, the breed was remarkable for possessing quite opposite characteristics from indigenous fighters like the Old English Game Fowl. Queen Vic is said to have adored them; she and her visitors found their huge, round bodies and genial temperaments absolutely beguiling. Perhaps they saw in them something as wholesome as the royal matriarch herself, with their ample rumps and ankles reassuringly hidden away under feathered legs.

Soon enough, people were searching all over China for similar novelties. The little Silkie is likely to have arrived during this period, as did the handsome, black Croad Langshan, launched in1872 by a Major FT Croad who said he had discovered it in the Langshan district, north of the Yangtse-Kiang River.

Meanwhile, back in the US, breeders crossed the Shanghai with Grey Chittagongs from India to create another exotic with an Indian name – the Brahma (from the 4-headed Hindu god of creation, via their sacred river the Brahmaputra). But it wasn't just exoticism they were after; breeders throughout the land were inspired to experiment with crossing these meaty Asian breeds with already established layers to find the perfect utility bird.

It is these sorts of combinations, crossing Asian imports with older ones (like the Italian Leghorn) that caused *Gallus gallus domesticus* greater evolutionary change than in the whole of her previous existence. Well done, you Victorians; I call that a triumph for East meets West.

Sadly, the forces of industrialisation ran counter to those of cultural exchange. It was in the Great British chicken psyche: having spent thousands of years watching them kill one another, within a century of banning the cockpit they were driving them into factories. All those colourful new breeds were all-too rapidly distilled into the egg and meat machines of today.

So here we are. The vast majority of chickens are now commercial hybrids. Contemporary geneticists like Hans Cheng may be concerned about these chickens' inability to cope with unforeseen circumstances, but his predecessors didn't think like that. Like their colleagues working for the car companies, they were doing a great job facilitating productivity. And when a disease like fowl pest happened to hamper their progress, they turned to the pharmaceutical companies who kindly provided the vaccines. It is much quicker and cheaper to spray a vaccine than it is to breed immunity over generations.

'But what if …?', cries the twenty-first century scientist. What if some pathogen arrives for which the manufacturers have not yet found a vaccine? It only takes one, and suddenly their equivalent to the Formula 1 racing car is crashing straight into one almighty chicken pile-up.

Step forward the Poultry Club of Great Britain. Established in 1877, the Poultry Club has a long tradition with a major and pivotal role in the conservation and preservation of pre-industrial stock. Amateurs all over the country keep all sorts of interesting breeds; for a modest membership fee they attend meetings at their local clubs, exchanging tips or even chickens. Many of them enter competitions up and down the country where their chooks are closely examined by expert Poultry Club judges. These people are potentially the keepers of that pool of rare genes that has disappeared from the commercial sector.

Every winter, the Club holds its annual shindig at the National Agricultural Centre in the Midlands. Over one weekend, breeders from all over the UK gather to show prime examples of traditional pure breeds and their eggs. At the opening in 1973, they hosted 1,367 entries; gradually they attracted more and more interest until, in November 2009, the number of entrants reached four times that. It was only just over an hour's drive away from where we live – an excellent family outing for a rainy Sunday.

As we followed the signs towards the entrance, through the warehouse doors came a screeching and a yelling – the sound of a thousand cocks asserting their territory. Even five-year-old Beatrice, who has attended many a children's party in her time, was shocked by the cacophony.

Inside stood rows and rows of cages with perfectly unsoiled wood-chippings on the bottom, a little pot of water and one of mixed grain attached to the door. Each cage contained a beauty contestant, his or her pristine winter plumage impressively displayed. Forget the home-keepers' rule about not washing your hen; these birds wouldn't contemplate going anywhere without a major wash 'n' blow-dry. The Great British Orpingtons were especially impressive: rows of caramel coloured giants, their plumage blown into the biggest bouffant you ever saw. Apparently the Queen Mother had several prize winners; of course, they couldn't be expected to lay much – far too posh to push.

Leaving Jay and the girls to gawp at the Orps, I went in search of the utility breeds. The row labelled 'Maran' was modest: just a few birds, all with Dark Cuckoo markings. Maybe I could arrange to borrow one of the cocks for a couple of weeks.

As I arrived, a shy young man was carrying one to the main table; how obedient that bird looked – just right for joining a trio in my backyard. His proud owner drew out the curve of his tail with his fingers; the cock tossed his red-hot comb and flashed a look at me. I flashed back. It was an oddly pleasurable sensation, flirting with a chicken.

An older chap in a flat cap was busily inserting green cards in the doors of the cages – 'I want everyone to see how I judged them,' he told me. 'They've already got their rosettes, but this gives them an idea of the detail.'

'Do you judge them for usefulness?' I asked.

'What do you mean?'

'Marans are meant to be utility birds, aren't they?'

'Well, yes – you'll see the way we judge them from the card.'

I made my way over to an empty cage with the first prize rosette on it. The winning bird had got 15 out of 15 for 'colour and marking', 5 out of 5 for 'legs and feet' but only 34 out of 40 for 'carriage and table merit'.

Oh dear. It seemed that even the best Maran in the country scored better on chicken beauty that on chicken dinner. It occurred to me that if a commercial hybrid was a racing car, then this star would be a vintage Bentley – beautiful but useless.

As I turned towards the exit, I spotted an interesting sign on top of the next-door row – 'Utility'. A man in official Poultry Club overalls was making his way past.

'Excuse me,' I interjected. 'Do you know anything about these?'

'I'm not sure I do,' he said, peering into the cages. Again, the row was short – not more than a dozen birds in total. I recognised a Barnevelder: a Dutch breed imported into the UK in the 1920s. I loved the look of him – his laced plumage, each feather with rings of greenish black on a nut-brown background. The female lays caramel brown eggs (at least 200 a year, they say).

Next door was a Plymouth Rock, her owner standing beside the cage.

'How come you've got a bird in this class?' I enquired.

'I didn't think she was in the running for her breed class, so I tried this one,' he said.

'And how do they judge her?' I asked, realising all of a sudden that it is much easier to judge a hen by her looks than by her productivity.

'They measure the pelvic bone,' said the man, raising his right hand with the fingers held tight together. 'Three fingers' width and she's a good layer.'

'And the meat?'

'That'll be the cockerels,' he said.

Come to think of it, that Barnevelder did have nice fat thighs.

'So why's it such a small class?'

'It only started a couple of years ago – Prince Charles is Patron of the Poultry Club, and he's quite serious about food, isn't he!'

Well, hurrah for Prince Charles. His grandmother may have gone for looks over productivity, but now he was making up for it.

This was exactly what I had been looking for, neither Formula 1 nor Bentley – the equivalent of a really reliable estate car.

I took the Plymouth Rock man's details and said I'd be in touch. It was time to find the rest of the family.

Jay had taken the girls down the end to the sales section. Beatrice was finding the competitive atmosphere all too much and had started to join in with the crowing. She was standing in front of a cage containing two large snowballs with blobs of black rubber on the top. Eventually, I managed to comprehend that she had an urgent need to take these things home; she wanted them for her Christmas present. I peered at the label – White Silkies.

They did look odd, with charcoal skin under white feathers so fragile they resembled that autumnal fuzz you find in country hedgerows – old man's beard. Come to think of it, the duo Bea desired looked more like Santas than snowballs. I knew they were one of the oldest breeds of hen: during his thirteenth century travels in Asia, Marco Polo wrote of a furry chicken that might well have been a Silkie. According to Hans Cheng and his team, it still has a full 100 per cent of the genes a chicken is meant to possess.

'Mummy,' announced Ellie, gravely. 'When I was five I got Snowy for Christmas.'

'But those aren't bunnies …' I started to protest. I had a vague feeling that Silkies were indeed once thought to be a cross between a rabbit and a hen … From Bea's point of view, that was exactly what they were. Bunny-chicks.

After all my confusions about keeping hens as pets, these were nothing but – terribly cute and easy to tame, apparently they don't damage the garden like other chickens; their life expectancy is a good nine years. The Silkie is neither utility bird nor hybrid, her flesh being unappetisingly dark and her egg-count no more than four a week. Her laying season starts over Christmas and finishes during the summer. The reason for this is uncertain – perhaps it is an acute sensitivity to light, stimulating her to start laying the moment the days get longer and to stop the moment they shorten. Or else, she stops because her Shanghai-original genes tell her the monsoon season has arrived and it is time to concentrate on keeping dry.

Her lack of meat and egg producing attributes doesn't mean she isn't useful: the Silkie's *raison d'être* is to bring up a family. Indeed, some say that the mere sight of an egg gets her mothering instincts going. Which makes me think how very unnatural our modern hybrids are. In the old days, maternal dedication was exactly what hens were celebrated for. The best-known allusion I can think of is in the Gospel according to St Matthew (Matthew 23:37), when Christ uses the hen as an example of selfless dedication to other people's children. She is the absolute opposite of the religious leaders with their brutality and their hypocrisy. At the high point of his diatribe, he cries – '*O Jerusalem, Jerusalem, you who kill the prophets and stone those sent to you, how often I have longed to gather your children together, as a hen gathers her chicks under her wings, but you were not willing.*'

Standing in front of the Silkies' cage, I was starting to think of another well-known piece of avian imagery – two birds with one stone. If I were to give in to Bea's pleas for a cuddly Christmas present, she would get the most adorable pets and I would get a couple of potential foster mothers for my allotment flock. I had been worried about my American Utility hens not becoming broody for their fertilised eggs, but, from her reputation, this

Chinese bird would always be there for them. A simple alternative to the incubator, the lamp and all those gizmos: just one fluffy bottom and a life of selfless dedication.

It's lovely how some breeds are prepared to look after foreign birds' babies. They are so satisfied by the experience of sitting and nurturing that they don't notice the breed. Come to think of it, the Orps with their huge bottoms have similar inclinations. Thank God Bea hadn't fallen in love with a couple of them – we would never have got them in through the front door.

A penchant for fostering was something we knew a bit about through ten-year-old Annie and her bantams.

One day Annie went out to the hen house to fetch the eggs and found Pong the Pekin refusing to leave the nesting box. When she tried to put her hand under her, the hen gave her an angry peck. That was weird, Annie thought, because Pong was usually so gentle. Her mum said she must be broody.

There was a moment of panic. Only now did the family realise that having Brendan the petrol-black cock meant all their eggs were fertilised and ready for a broody. All of a sudden, with no forward planning, they had shifted from simple egg-production straight into parenting. They borrowed a little broody house from a friend and moved Pong in there with all the eggs she was sitting on (five brown and one white). A week later, they found that she had pecked the white egg; red veins were leaking out on to the straw. Annie had an idea about this – she thought Pong was worried that this chick was going to be different, like its egg. Maybe she thought the others would bully it, so she got rid of it. Interestingly, this was the only one she could possibly have laid herself; the brown ones were from their hybrids and could only ever become her foster-chicks.

Anyway, Annie really liked looking after Pong while she was sitting on her foster-eggs. Every morning before school she took her a handful of corn and fed it to her while she sat on her nest. She ate and she ate – half a cup or more, sometimes.

After three weeks, Annie's brother was out one day and he had a look at the eggs: there was a hole in one of them. He came running to get her and when she had a look she could see a little yellow eye peaking through the hole. She thought maybe the chick had got stuck, so decided to take a bit of shell off. That must have helped, because in just a few minutes she was out, all wet and pale; in a flash she was hidden beneath her foster-mum, but not before Annie had decided to call her Snow White.

Unfortunately, none of the other eggs hatched, which may well be because the children had interfered in the process. Annie said she didn't mind – she was sure Snow White would always have been her favourite. She used to run away from everyone, but with Annie she was docile; she would let her stroke her and carry her about, which Annie was convinced was because she had been there at the birth.

Until Snow White started to crow. At which point he was rechristened Rocky and left to strut about the garden, a perfect compliment to his dad.

In our case, if this Silkie plan went ahead I intended to prevent my children helping with the birth. Young Beatrice would have to keep well away from her broody Silkie when the eggs were hatching. And if she went and fell in love with a boy chick, I would have to hark on local traditions of half a century ago, when Dianne and her siblings were forced to part with their beloved, and later to eat him.

And to speed things up, I might decide to circumvent the whole *ménage à trois* insemination thing. The gallinaceous equivalent of

IVF has a long and reliable history: you simply get hold of some fertilised eggs from a reputable hatchery or friend and pop them underneath your already broody Silkie. There are loads available on eBay – they should arrive safely in the post, as long as they didn't have too far to travel and the postman didn't shake his parcels around too much. It would be sad not to have something as handsome as that Maran to stay, but probably a lot less hassle. And soon enough, we might have our own such bird for the meat 'n' egg plot …

And just at that moment, just as I was reaching the decision that Silkies were indeed my family's future, who should arrive at my shoulder, as if from nowhere, but John Marfleet, our piano tuner. It was such an amazing coincidence, it could only happen in real life.

'Like them Silkies do you?' said John Marfleet. John breeds bantams.

'They make good pets?'

'I'd say so. Broodies is what people use them for … nice looking pair. How much are they?'

He shifted his specs and peered at the price tag.

'Fifty quid! You're not paying that!'

'I was thinking of getting them for the girls for Christmas.'

'You know, I've got some Silkies coming along.'

'Full size?'

''Course! 'Lot cheaper than these ones, I'm telling you.'

'How much, John?'

'I dunno – I'll probably let you have them for nothing. 'Round nine weeks, when their Mum comes off of 'em.'

'Will you know which are the cockerels by then?'

'Mmm. Not sure – I've never been that good at telling the Silkies … We'll have to see.'

I looked down at my daughters whose faces were wide with joy. John Marfleet had just made their Christmas.

On our way home through the winter darkness, Ellie and Bea chatter about what we are going to call their new pets. *Fluffy and Snowball ... Barley and Corn ... Acorn and Popcorn ... Lola and Mabel ...* The possibilities are endless. Jay is just happy that this year's Christmas present originates in China but is not the usual plastic variety. And I am busy fantasising about next year's brood.

If we want chicks by Easter, we'll need one of the Silkies to be broody by March. She'll need fattening up because once in charge of the eggs she is going to spend a good three weeks glued to them. I will also need to give her a good going over for lice and mites; the last thing an expectant mother wants is an infestation.

I'm going to need a peaceful place for her to nest. A tough cardboard box with holes will do, with lots of woodchips and straw on the bottom to keep things comfortable. If the weather is still chilly, I'll give her a corner at the end of the kitchen, behind the chest of drawers; if it is warm, I could probably give my rabbit cage one last reincarnation – the sleeping compartment is perfect for nesting and the little run area will be useful for the chicks when they make their first forays abroad.

John will advise me how many eggs I need to get – they say nine for an average sized hen; seven might be a good number for a Silkie, and a lucky one for me, just slightly superstitious about my new venture.

We'll transfer the broody to her box (or cage) and once we're sure she is there for the duration, I can ease the clutch under her one by one. Then it will just be a case of waiting.

And what about feeding her? I shall have to feel my way on that one. Probably my children will be like Annie and want to hand-feed her, but John says not to bother; his broodies have always been so fixated on their job that he just leaves them alone for the full three weeks' incubation. All they get is a little trough of water that they can reach through the bars of his broody coop. My pre-war mentor, Rosslyn Mannering, advises intervention. He says take the hen off the eggs once a day; if after 10 minutes or so she still has not defecated, Mannering says you should scoop her up by the wings, hold her for a moment suspended about a metre from the ground and then suddenly drop her. He says not to worry – it really is no worse than her hopping off the perch, and 'the expansion and contraction of the muscles engendered seldom fails to produce the desired result.'

Oh, and if the children promise to be really careful, Mannering says while she is out pooing they can moisten the eggs with a little tepid water. Day 19 is when this is particularly important, in preparation for hatching; John says his broodies did the job themselves, swooshing water with their beaks from the trough into their nest.

Come Day 21, fingers crossed, all seven little darlings will appear under their foster Mummy's tummy. And soon enough all our friends and relations will be crowding in to welcome them.

The traffic is dense; lights multiply in the raindrops across the windscreen – red ones streaming forward and white ones against us. Even on a Sunday, there is not room enough for all the cars. Drivers veer between lanes, nudging in and out of one another's path in order to get that little way ahead and closer to home. There are just too many of us; too many humans needing too many cars; too many chickens and too many eggs. As every year our number increases, how are we going to cope?

Struck by the impossibility of the twenty-first century lifestyle, I am starting to have bold ambitions for my new chicks. I have this idea that maybe, once we are up and breeding, we will get a cock for the Silkies and breed from them. What I am thinking is that when that 'disease shock' comes along and wipes out the hybrids, our birds will call on their 100 per cent genetic reserves and survive. Suddenly they will be the only chickens on the planet – unappetisingly black, with modest laying abilities. All that pappy white flesh and rows of pseudo-rustic brown eggs will be gone; and, instead, a rare and cherished sight will be a little Chinese chicken, pecking around somebody's back garden.

Glossary of Chicken Terms

Alektorophobia – a fear of chickens.

Bantam – a little chicken, often a miniature version of a larger breed, though bantam breeds exist in their own right, for example the Pekin (given the name of the Chinese capital city by the Victorians, but probably originating in Java like all true mini-hens).

Broiler – a chicken bred for its meat.

Chicken – The word 'chicken' used to describe not the adult but the baby – hence the term 'she's no spring chicken' and that seeming tautology of a pub, 'The Hen and Chickens'. 'Chick' is simply a shortened version, introduced (it is thought) around the fourteenth century.

Chook – All over the internet you come across this generic term covering all ages of chicken and both sexes; the word is said to originate in the Antipodes.

Cock – a sexually mature male chicken.

Cockerel – strictly speaking this word refers only to a male chicken under a year old. In the UK, Canada and Australia many people coyly use the term for the mature male.

Dual Purpose – breeds of chicken that used to provide both meat and eggs.

Gallus – Apart from the domestic chicken (in all its forms), this genus of bird includes four varieties of Junglefowl living wild in South and South East Asia.

Gallus gallus domesticus – the Latin name for a domestic chicken.

Gallinaceous–belonging to the genus *Gallus.*

Hen – a mature female chicken, probably over a year old.

Pullet – a female under a year old.

POL – a pullet 'at the point of lay' (normally between 16 and 20 weeks old).

Rooster – the American word for cock.

Utility bird – same as 'dual purpose' – 'utility' as in something useful.

Acknowledgments

Thank you to all my friends who helped create this book. It has been such fun talking chickens with you, and getting your ideas down on paper. I am only sorry there was not room for all your names and all your stories.

I would also like to thank the following organisations and individuals who went out of their way to share their expertise:

Phil Brooke at Compassion in World Farming
Anna Bassett from the Soil Association
Elm Farm Research Centre
Mike Gooding and colleagues at FAI
Greger Larson, dept. Archaeology, Durham University
Jane Howorth, BHWT
Alice Clark and Mark Cooper from RSPCA's Freedom Foods
Humane Slaughter Association
National Society of Allotment and Leisure Gardeners Ltd.
The Poultry Club of GB
Rabbi Eli Brackman
Rev. Andrew McKearney

Rev. David Barton

Fiona Tomley and colleagues at the Institute for Animal Health, Compton, Berkshire.

As for the publishing bit – thanks to Lisa Darnell for so dexterously overseeing the project, thanks to Nigel Wilcockson for his imagination and encouragement, and to Vanessa Neuling for fine-tuning.

Finally, a special thank you to my family – to my father and father-in-law for rigorous attention to my prose, and to Jay and the girls for putting up with yet another encroachment on family life.